Coming Clean

American and Comparative Environmental Policy
Sheldon Kamieniecki and Michael E. Kraft, series editors

A complete list of the books published in the American and Comparative Environmental Policy series appears at the back of this book.

Coming Clean

Information Disclosure and Environmental
Performance

Michael E. Kraft, Mark Stephan, and Troy D. Abel

The MIT Press
Cambridge, Massachusetts
London, England

For information about special quantity discounts, please email special_sales@ mitpress.mit.edu

This book was set in Sabon by Toppan Best-set Premedia Limited. Printed and bound in the United States of America.

Library of Congress Cataloging-in-Publication Data

Kraft, Michael E.
Coming clean : information disclosure and environmental performance / Michael E. Kraft, Mark Stephan, and Troy D. Abel.
    p. cm. — (American and comparative environmental policy)
Includes bibliographical references and index.
ISBN 978-0-262-01495-3 (hardcover : alk. paper) — ISBN 978-0-262-51557-3 (pbk. : alk. paper) 1. Environmental policy—United States—Decision making. 2. Toxics Release Inventory Program (U.S.) 3. Environmental reporting—United States. 4. Disclosure of information—United States. I. Stephan, Mark. II. Abel, Troy D. III. Title.
GE180.K725 2011
333.71′4—dc22
                                                                            2010015784

10   9   8   7   6   5   4   3   2   1

Contents

Series Foreword   vii
Preface and Acknowledgments   xi

1 Information Disclosure and Environmental Performance   1

2 How Does Information Disclosure Work?   23

3 Reducing Toxic Releases and Community Risks   53

4 States of Green: Regional Variations in Environmental Performance   83

5 Facility-Level Perspectives on the TRI and Environmental Performance   119

6 Environmental Leaders and Laggards: Explaining Performance   153

7 Conclusions and Policy Implications   177

Notes   203
References   221
Index   241

Series Foreword

The numerous illegal activities by leaders of some of America's largest corporations at the turn of the century undermined the trust and confidence the public had in private firms and the ability of corporations to promote social responsibility was understandably questioned. More recently, the failure of large banks and investment firms, the enormous bonuses their top executives continue to receive, and the Bernard Madoff scandal have also led to increased cynicism concerning the true willingness of companies to promote the public good. Clearly, public trust in American business has significantly eroded over the past decade as a result of the unparalleled charges leveled against once highly respected companies and their executives and the flood of media coverage that has accompanied government action.

The decline in public confidence in corporate America has the potential to spill over into important policy areas, such as environmental policy, and to color government actions in those issue areas. Despite these incidents, firms continue to ask the government to end direct command-and-control regulation and trust them to select the most cost-effective methods and equipment to abate pollution and manage natural resources. Business leaders believe this can be accomplished by instituting market-based mechanisms and voluntary action programs. Many large firms now support efforts to improve environmental quality and are willing to change certain practices in order to achieve that end. Environmentalists, however, continue to complain that corporations are not sincere, only care about profits, and will not cease polluting or wasting natural resources unless forced to do so by government.

This outstanding book written by Michael Kraft, Mark Stephan, and Troy Abel investigates how the process of information disclosure, an approach that requires minimal direct government intervention, works and the effects it has on environmental performance within companies and on the localities within which their facilities are located. The theory behind this strategy is that firms will consciously make an effort to reduce pollution if they know ahead of time that they will have to disclose publicly the amount of pollution they generate. Firms will do everything they can to avoid this kind of negative attention and publicity.

Specifically, this book analyzes the federal Toxics Release Inventory (TRI) program to explore how industrial facilities throughout the country have tried "to come clean" in two interrelated ways: to disclose key information to the public about the management and release of toxic chemicals and the health and environmental risks they pose to surrounding communities, and to use that information to reduce the amount of pollution being generated. The investigation seeks to find out how much success the TRI program has achieved in both respects. Among the major research questions the book addresses are: How has the disclosure of such information made a difference within the community and state, and how has compliance with the TRI program affected company management of toxic chemicals? The book also seeks to uncover how changes in the control of toxic chemicals have come about, that is, the mechanisms by which the collection and disclosure of information affects community and corporate decision making. Finally, the authors examine what the program's accomplishments suggest about the potential for information disclosure as an environmental policy strategy.

Kraft et al. focus on the TRI program because it is an excellent example of what most observers refer to as a federal, non-regulatory environmental policy. The TRI imposes mandatory reporting requirements on affected facilities and, in that respect, it shares at least one characteristic of other environmental policies that are considered to be regulatory. Yet the program does not include the regulatory apparatus of command and control that accompanies conventional environmental policies such as the Clean Air Act and Clean Water Act. Because it is so different from traditional environmental regulatory approaches, the effects it has had since its adoption in 1986 speak to the broader question of how information disclosure might be used to achieve environmental objectives at a time when confidence in industry, in general, and in conventional command-and-control regulation, in particular, has weakened. Policy analysts across the political spectrum have called for consideration of a range of new approaches to environmental policy, from market incentives to flexible regulation. This ongoing debate highlights the need to learn more about how well present policies are working and what might be expected of new strategies that are widely supported today.

Kraft et al. employ a variety of quantitative and qualitative research techniques to address these questions, including extensive analysis of the TRI database itself and consideration of both the quantity of chemical releases and the public health risks associated with those releases. They also survey a national sample of corporate TRI officials, federal and state government policymakers whose work includes oversight of the TRI program, and emergency management personnel at the community level whose responsibilities include at least some elements of local chemical management. Lastly, the researchers employ integrative and illustrative case studies of facilities and localities across the nation to learn more about how changes in industrial operations work at the local level, and the driving forces that bring them about. Some of the findings they report in the book run counter to public understanding of how the TRI program or information disclosure works. Thus, this investigation has important implications for any attempt to redesign the TRI program itself or to consider other information disclosure policies.

The book includes seven well-written, highly integrated chapters. The first two chapters address information disclosure and its relationship to environmental performance, with an in-depth explanation about how information disclosure works. Succeeding chapters focus on the reduction of toxic releases and regional and facility variations in environmental performance and present an analysis of variations in environmental compliance and performance. At the end of the volume the authors expertly tease out several important conclusions and policy implications from their research. Kraft et al. provide an insightful and critical investigation of the TRI program, and they offer a number of promising new directions for future research.

The book illustrates well the goals of the MIT Press series in American and Comparative Environmental Policy. We encourage work that examines a broad range of environmental policy issues. We are particularly interested in volumes that incorporate interdisciplinary research and focus on the linkages between public policy and environmental problems and issues both within the United States and in cross-national settings. We welcome contributions that analyze the policy dimensions of relationships between humans and the environment from either a theoretical or empirical perspective. At a time when environmental policies are increasingly seen as controversial and new approaches are being implemented widely, we especially encourage studies that assess policy successes and failures, evaluate new institutional arrangements and policy tools, and clarify new directions for environmental politics and policy. The books in this series are written for a wide audience that includes academics, policymakers, environmental scientists and professionals, business and labor leaders, environmental activists, and students concerned with environmental issues. We hope they contribute to public understanding of environmental problems, issues, and policies of concern today and also suggest promising actions for the future.

Sheldon Kamieniecki, University of California, Santa Cruz
American and Comparative Environmental Policy Series Co-Editor

Preface and Acknowledgments

Business today clearly takes its environmental performance far more seriously than it did over the past several decades. Corporate managers increasingly understand that a positive environmental record matters to the public and to shareholders, and as a consequence many tout the green credentials of their companies whenever the opportunity presents itself. Industrial facilities seem to have been affected by these changes at least as much as have as other businesses; they are much more attentive than previously to the need to use cleaner methods of production, decrease manufacturing waste, improve energy efficiency, reduce greenhouse gas emissions, and incorporate sustainability goals into their operation. In this book we focus on one aspect of the corporate greening: the disclosure of information about toxic chemical releases that documents environmental performance and also meets public expectations for transparency and social responsibility.

In particular, we examine the federal Toxics Release Inventory (TRI) program to explore how industrial facilities around the country have tried to come clean in two interrelated ways. One is to disclose key information to the public about the management and release of toxic chemicals and the health and environmental risks they pose to surrounding communities. The other is to use that information to clean up or "green" their activities. We want to know how much the TRI program has achieved in both respects. How has the disclosure of such information made a difference within the community and state, and how has compliance with the TRI program affected company management of toxic chemicals? We also want to know how changes in the management of toxic chemicals have come about, that is, the mechanisms by which the collection and disclosure of information affects community and corporate decision making. And we want to know what the program's accomplishments say about the potential for information disclosure as an environmental policy strategy.

We focus on the TRI program because it is the premier example of what most observers refer to as a federal, nonregulatory environmental policy. The TRI imposes mandatory reporting requirements on affected facilities and in that respect it shares at least one characteristic of other environmental policies that are considered to be regulatory. Yet the program does not include the regulatory apparatus of command and control that accompanies conventional environmental policies such as the Clean Air Act and Clean Water Act. Because it is so different from environmental regulatory policies, the effects it has had since its adoption in 1986 speak to the broader question of how information disclosure might be used to achieve environmental goals at a time when confidence in conventional command-and-control regulation has weakened. Commentators and policy analysts across the political spectrum have called for consideration of a range of new approaches to environmental policy, from market incentives to flexible regulation. This ongoing debate highlights the need to learn more about how well present policies are working and what might be expected of new approaches that are widely endorsed today.

Relatively few studies offer solid evidence about the effectiveness, efficiency, or equity of either long-standing regulatory policies or the newer approaches so much discussed in the 1990s and 2000s. In the case of the TRI program, scholars have explored its origins, operations, and some of its impacts with considerable insight. Our purpose in this book is different. We are interested in the mechanisms through which the program affects industry and communities, and whether and how it has helped to reduce the release of toxic chemicals and their risks to communities across the nation.

We also use the TRI program as a lens through which we can examine the environmental performance dilemma faced by government and industry as both strive to make the industrial footprint consistent with long-term goals of sustainability. The dilemma is that with conventional command-and-control regulation, government and business may pursue strategies that deliver less than optimal results. Yet there are ways to create win-win solutions where environmental performance can be improved through the use of hybrid policies that combine regulation and information disclosure, among other policy tools.

The TRI data are not without some problems of reliability, as critics have long observed. Yet used cautiously, we believe that the TRI database provides a unique and valuable opportunity to examine and compare corporate environmental performance over time and across the range of industrial facilities, and to ask whether and how the TRI spurs new environmental management strategies among government agencies, non-governmental organizations, corporations, and communities. We employ a diversity of quantitative and qualitative research methods to address these questions, including extensive analysis of the TRI database itself and consideration of both the quantity of chemical releases and the public health risks associated with those releases. We also surveyed a national sample of corporate TRI officials, federal and state government officials whose work includes oversight of the TRI program, and emergency management personnel at the community level whose responsibilities include at least some aspects of local chemical management. Finally we use integrative and illustrative case studies of facilities and communities across the nation to learn more about how changes in industrial operations work at the "street level," and the driving forces that bring them about.

Some of the findings we report here run counter to public understanding of how the TRI program or information disclosure works. Thus we believe the analysis has important implications for any attempt to redesign the TRI program itself or to consider other information disclosure policies. Information disclosure alone is not the panacea that many proponents of TRI-like programs assert it is. However, it is an important piece of a multiple-method approach to fostering safer and cleaner manufacturing. Our findings also illuminate specific paths that government and industry can take to achieve safer and cleaner industrial operations.

As we note at several points in the chapters that follow, we hope that others will choose to address with further research at least some of the questions that we explore in this book. To that end, we have made the full TRI database that we employ and all of our survey results available through the Data Preservation Alliance for the Social Sciences project of the Inter-University Consortium for Political and Social Research at the University of Michigan. For one year following publication of the book, we also will keep the Information Disclosure and Environmental Decision Making Web site at the University of Wisconsin-Green Bay active. All of the questionnaires, survey results, and the major papers that we have presented at scholarly and government meetings are archived at that site: www.uwgb.edu/idedm.

We are indebted to many whose support and assistance made this book possible. The research was supported by the National Science Foundation (NSF) under grant number 0306492, Information Disclosure and Environmental Decision Making. We are particularly grateful to Robert O'Connor, co-director of the Decision, Risk, and Management Science Program at NSF, for his continued faith in and support for the project. Of course, any opinions, findings, conclusions or recommendations expressed in this book are those of the authors and do not necessarily reflect the views of the National Science Foundation.

As always in research of this kind, we received substantial logistical support from our respective campuses: the University of Wisconsin-Green Bay, Washington State University-Vancouver, and Western Washington University. Throughout the project we were fortunate to have worked with a number of undergraduate and graduate research assistants, including Ellen Rogers, Grant Johnson, Timothy Larsen, Kristin Neveau, Jennifer Faubert, Jonathan Schubbe, Erin Busscher, Alex Vasiliev, Patricia Robert, Lauren Miller, and Eric Ryer. Derek Nejedlo and Paula Ganyard at the UW-Green Bay Learning Technology Center assisted with making the national surveys available online and with developing the project's Web site. Chris Terrien, Lidia Nonn, and Jeff Selner assisted with project financial management at UW-Green Bay, while Peggy Bowe, Dotty Morlan, Ginny Taylor, and Marie Loudermilk performed similar tasks at WSU Vancouver.

A number of our colleagues offered helpful advice along the way on project papers presented at conferences and in other venues, including Barry Rabe, Sheldon Kamieniecki, Daniel Fiorino, Mark Lubell, Paul Sabatier, Matt Potoski, Dorothy Daley, Don Grant, Tom Beierle, Archon Fung, and Leigh Raymond, among others. We are grateful as well for the comments and suggestions offered by the anonymous reviewers for MIT Press. We owe special thanks to Rose Krebill-Prather and the Social and Economic Sciences Research Center at WSU, whose aid in development of the survey instruments was invaluable. We especially want to thank the corporate and governmental respondents to our national surveys for taking valuable time to answer our questions about the TRI program and their experience with it. Naturally, we assume responsibility for any errors or omissions that somehow have escaped our notice during the writing, editing, and production of the book.

1

Information Disclosure and Environmental Performance

By all accounts, industrial corporations today take their environmental performance seriously. Increasingly, corporate managers believe that people care about the way companies affect the environment, and they recognize the need to show due regard for the health of the communities in which they have a facility. As a result, many companies of widely varying size and across every industrial sector regularly tout their green credentials in print advertisements and television commercials. Many of them also highlight their environmental achievements and aspirations on company Web sites and in annual reports to shareholders.

The popular press has reported favorably on these developments, often celebrating the greening of industry in general and the environmental accomplishments of companies whose green credentials have been seen as particularly impressive, including General Electric, S.C. Johnson, Johnson Controls, Duke Energy, DuPont, and Wal-Mart. At the community and state levels, the media similarly report on environmental milestones of local businesses, such as the use of cleaner methods of production, reduction in hazardous waste that is produced, improved energy efficiency, reduced greenhouse gas emissions, and the creation of "green jobs" through the manufacture or installation of products such as wind turbines and solar photovoltaic panels. Skeptics are quick to charge that much of the new green promotion is "greenwashing," or a corporate public relations gimmick, while business as usual continues. Yet recent years have brought a palpable shift in corporate environmental behavior that merits serious consideration.[1]

Consistent with these new beliefs and commitments, corporations release a great deal of technical information that documents their environmental performance and meets new public expectations for transparency and social responsibility on the part of corporate managers. At least some of that information can help to inform surrounding communities of public health and other risks associated with the activities of manufacturing facilities. By coming clean in these ways, the facilities' managers not only acknowledge the pollution associated with their manufacturing activities but stand to learn from the process, possibly finding new ways to reduce their environmental footprints while improving their economic bottom line. In this book we seek to understand just how this process of information disclosure works and the effects that it has on environmental performance within companies and on the communities within which their facilities are located.

The potential for information disclosure or provision to achieve these lofty goals is of special interest at a time when public and corporate confidence in conventional regulation has waned.[2] For nearly four decades, environmental protection policies in the United States have required that industrial facilities meet certain targets for release of chemicals to the air, water, and land. Many of these policies have mandated the use of specific technologies and forced industry to achieve the maximum improvement possible, and more than a few have specified in exceptional detail which chemicals were to be managed and how. The goal was to provide a degree of certainty that businesses would indeed achieve the new standards and that the implementing agencies, particularly the U.S. Environmental Protection Agency (EPA), would not have so much discretionary authority that they could thwart the will of Congress.

By one recent account, some 15,000 pages of federal regulations are needed to provide instructions for companies and other entities covered by the laws, and "an elaborate system of reporting, inspections, and penalties exists to make people follow the rules" (Fiorino 2006, 1). This description applies to all of the major national environmental protection policies: the Clean Air Act, Clean Water Act, Resource Conservation and Recovery Act, Safe Drinking Water Act, Toxic Substances Control Act, Comprehensive Environmental Response, Compensation, and Liability Act (Superfund), and Federal Insecticide, Fungicide, and Rodenticide Act—and their later amendments. In some cases, Congress made those amendments, such as the 1984 revision of the Resource Conservation and Recovery Act, even more detailed and demanding than the original statutes because it grew increasingly distrustful of the EPA (and the White House) and sought to ensure that the agency would proceed on course. The collective reach of the laws is astonishing and their implementation is a daunting task. The total number of facilities whose environmental performance has been regulated by the federal government or the states in the 1990s and 2000s includes an estimated 40,000 stationary air sources, 90,000 facilities with water permits (which cover about a half-million sources), over 425,000 hazardous waste facilities, 400,000 underground injection wells, and 173,000 drinking water systems (U.S. EPA 1999).

As might be expected, setting and enforcing regulations to carry out and comply with the core federal pollution control statutes on this scale is not cheap. In recent years the nation has likely spent more than $200 billion annually; most of that money, nearly 60 percent, comes from corporations seeking to meet their regulatory obligations.[3] The cumulative expenditures since the 1970s obviously are very large. At the same time, there is little question that the dominant command-and-control regulatory policies of the past four decades have produced real and important improvements in environmental quality and in public health, the values of which have exceeded the costs of regulation. For example, the EPA estimated that the cumulative benefits of clean air regulation between 1970 and 1990 ranged between $5.6 and 49.4 trillion, with a mean estimate of $22 trillion. In contrast, the direct compliance costs were estimated to be only $0.5 trillion (U.S. EPA 1997). Moreover, annual reports to Congress by the Office of Information and Regulatory Affairs regularly find that the value of benefits produced by new environmental protection regulations exceed the costs, often by wide margins.[4]

The evidence of improving environmental conditions also is clear and fairly well documented by the EPA and the states. For example, the nation's air and surface water are cleaner, drinking water is safer, hazardous chemicals are better managed, and the release of toxic chemicals to the environment has been significantly reduced. All of this has been achieved while the nation's economy, population, and energy use have grown substantially. The data on air quality are particularly striking and often cited as a major sign of such progress. The EPA reports that emissions of the principle pollutants controlled by the Clean Air Act decreased by 60 percent from 1970 to 2008 even while the nation's economy grew by 209 percent, the population rose by 44 percent, vehicle miles traveled increased by 163 percent, and overall energy consumption grew by 49 percent (U.S. EPA 2008). Although debate continues over precisely how best to measure changes in environmental conditions and how to document improvements systematically in light of significant gaps in data collection and reporting, this broad and impressive set of accomplishments is noteworthy. Some would add that had the EPA been more fully funded over the years and had the agency more aggressively enforced the laws, the environmental outcomes likely would have been even more striking (Eisner 2007; Kraft 2011; Portney and Stavins 2000).

Nonetheless, many critics today believe that continued progress in corporate environmental performance cannot be assured through reliance on the core statutes of the 1970s, no matter how well implementation goes. There are a number of reasons for this conclusion. First, most of the large sources of pollution have already been identified and controlled, and most major corporations are in compliance with existing regulations; thus there are important limitations on how much more can be easily achieved through the regulatory process. Second, as tougher standards are considered over time, the costs of compliance can rise substantially because of the higher marginal costs of improvement in environmental performance; thus substantial economic barriers to progress can arise. Third, many of the remaining sources of pollution are far less amenable to command-and-control regulation, and they call for different approaches. An example is surface water pollution, where the remaining problems are attributable largely to nonpoint sources, such as urban runoff and agriculture, which cannot easily be regulated. Much the same could be said about the millions of mobile sources of air pollution, such as passenger vehicles, trucks, and buses. Further regulatory requirements to improve gasoline formulation or the use of additional pollution control equipment may be less appropriate in the future than the provision of incentives for developing alternative technologies, including hybrid, electric, and fuel-cell powered vehicles.

Beyond these constraints are the long-standing criticisms directed at the environmental regulatory system itself. As one recent appraisal put it, the policies that have contributed to the notable environmental outcomes described just above have been widely viewed as "heavily bureaucratic, prescriptive, fragmented in purpose, and adversarial in nature" (Durant, Fiorino, and O'Leary 2004, 1). Businesses and other critics have long complained as well about the overall complexity and rigidity of rules and regulations, the high costs of compliance with policy requirements, the focus on remedial rather than preventive actions, the difficulty of using management strategies that cut across different environmental media, and the lack of incentives for companies to innovate or go beyond compliance with regulatory standards to achieve better environmental results (Davies and Masurek 1998; Eisner 2007; Fiorino 2006; Schoenbrod 2005). More recently, another important line of criticism has been advanced. This is the inability of the various statutes, singly or collectively, to steer the nation toward the essential long-term goal of sustainable development, which requires a far more comprehensive and integrated approach to environmental problem solving than possible with existing environmental laws (Mazmanian and Kraft 2009).

In addition, the EPA has long suffered from variable but limited public and policymaker support, which arguably is essential for the regulatory process to succeed. The American public clearly is concerned about pollution and has long favored tough regulation to protect its health. Yet generally people pay little attention to the EPA and its decisions, and they have little awareness of the substance of environmental policy. The issues are rarely salient enough to stimulate most people to become more informed or active, for example, to contact the EPA or the states on regulatory standard setting or enforcement actions (Guber and Bosso 2010).[5] Public action of this kind comes more often from the organized environmental community rather than from the citizens themselves. The business community, on the other hand, is highly attentive to such agency decisions, often is sharply critical of them, and may lobby intensely for less demanding and less costly regulations (Kraft and Kamieniecki 2007).

This is an impressively long list of significant weaknesses or failures in four decades of U.S. environmental policy. It is hardly a surprise, therefore, that at least since the early 1980s an expansive and varied, though often ill-defined, agenda for environmental policy reform has emerged and that policymakers, analysts, and scholars have advanced and discussed such reforms extensively (e.g., Dietz and Stern 2003; Durant, Fiorino, and O'Leary 2004; National Academy of Public Administration 1995 and 2000; Sexton, et al. 1999).

Sadly, despite all of the concern, critiques, deliberation, and occasional experiments with new approaches, very little has changed in the prevailing environmental policy regime, especially at the federal level. The first generation of environmental regulatory policies from the 1970s, with its many, well-documented flaws, largely continues in force in part because of persistent political stalemate over precisely what kinds of changes to make and uncertainty over who would gain or lose as a result. Business groups and political conservatives have favored one set of solutions, including greater use of market incentives and flexible regulation. Environmental organizations have feared that opening the core statutes for consideration of such fundamental changes risks losing many of the gains of the previous decades. Each has been powerful enough to block the other's policy proposals. The EPA itself has experimented with many different approaches to regulatory flexibility and voluntarism, particularly during the 1990s and early 2000s (Dietz 2003; Mazurek 2003). In the end, however, the agency found itself hobbled by existing statutes, congressional reluctance to grant it more discretion, and its own organizational culture, which has not accorded policy and administrative reform a high priority (Eisner 2007; Fiorino 2006; Marcus, Geffen, and Sexton 2002).

To be sure, one finds many important and often innovative policy changes at the state and local level, and indisputably significant elements of change in federal administrative rules and procedures, court decisions, and congressional funding actions even if Congress has remained mired in gridlock on the major statutes (Klyza and Sousa 2008; Kraft 2010; Vig and Kraft 2010). In a few striking cases, members of Congress have agreed on substantial legislative changes that incorporated elements of the reform agenda. The Clean Air Act Amendments of 1990, and especially the cap-and-trade program for control of acid rain, and the Food Quality Protection Act of 1996 (which modernized regulation of pesticides) are examples. Moreover, it is equally evident that many corporations have launched major environmental and sustainability initiatives on their own, undeterred by the failure of federal policymakers to chart the way (Esty and Winston 2006; Press 2007; Press and Mazmanian 2010). So while it is clear that this conversation over a new generation of environmental policy will continue for many years to come, the need for action has hardly gone unnoticed.

Much of the discussion about new directions in environmental policy has focused on the likely effectiveness, efficiency, or public and political acceptability of alternatives to federal command-and-control regulation. Many alternatives that have been identified, appraised to some degree, and endorsed by a diversity of policy actors. These include a plethora of voluntary initiatives by business and voluntary public-private partnerships (Potoski and Prakash 2009; Prakash and Potoski 2006); more frequent use of market incentives (Freeman 2006; Olmstead 2010); flexible regulation based on environmental results or performance (Fiorino 2004); greater involvement of citizens and other stakeholders in regulatory decision making, particularly through more open and collaborative processes often termed "civic environmentalism" (Abel and Stephan 2000; Agyeman and Angus 2003; John 2004); further decentralization of environmental responsibilities to the states and local or regional governments (Rabe 2010); and greater use of information disclosure (Hamilton 2005).

Critics of existing environmental policies suggest that in many different ways such new approaches can supplement, and perhaps eventually replace at least some of the command-and-control regulation now in place (Durant, Fiorino, and O'Leary 2004; Fiorino 2006; John 2004; Schoenbrod, Stewart, and Wyman 2009). They may well be correct, but often it is difficult to know with any certainty. This is in part because there have been relatively few careful assessments of how such approaches have worked in practice or what their potential may be for the future even if those that have been completed suggest the considerable value of such analysis (Borck and Coglianese 2009; Coglianese and Nash 2001, 2006a, 2006b; Dietz and Stern 2003; Harrington, Morgenstern, and Sterner 2004; Harrison 2003; Morgenstern and Pizer 2007; Wilbanks and Stern 2003). We hope our study of environmental information disclosure and its impacts on corporations and communities can speak to these concerns and also stimulate further inquiry into the promise of a new generation of environmental policy.

We are not so naïve to believe that information disclosure by itself, no matter how well designed and implemented, can work miracles. But we believe that it can be an important element in a comprehensive and multifaceted approach to environmental protection. Thus we want to understand its potential and limitations, and the factors that influence its success in different corporate, community, and governmental contexts. In designing this kind of study, we follow in the footsteps of a growing body of recent scholarship that also has sought to analyze new policy approaches through use of a rich variety of complementary methods to better understand their achievements and potential (Layzer 2008; Lubell 2004; Mazmanian and Kraft 2009; Prakash and Potoski 2006; Sabatier et al. 2005; Weber 2003; Weible and Sabatier 2009).

Information Disclosure Policies

Our study began with a focus on the federal Toxics Release Inventory (TRI), established by Congress in 1986, in part because the TRI was the first major federal environmental protection program based not on adversarial command-and-control approaches but rather on industry self-disclosure of environmental performance information. Even though many state agencies use the TRI data as part of their regulatory efforts— and the reporting of TRI data are mandatory for the affected facilities— it is nonetheless accurate to characterize the federal program as nonregulatory in its design and implementation. As our research unfolded, we expanded our investigation to consider environmental performance as measured by changes over time in TRI data. Although our analysis concentrates on TRI data, we believe the study's findings have implications for many other kinds of information disclosure policies as well as for other alternatives to regulation that continue to be debated. As for information disclosure policies themselves, they are found increasingly at all levels of government, and there is every reason to think that public demand for information about corporate and government actions will continue apace.[6]

Consider the variety of policy areas in which some form of information release is a central component (Weil, Fung, Graham, and Fagotto 2006; Weiss and Tschirhart 1994). Federal campaign finance reforms of the past several decades are at heart based on making public the contributions given to candidates for federal office and the sources; public knowledge of the sources of funding is thought to make elections more open, honest, and accountable. Following the scandalous actions in the early 2000s on the part of Enron, Tyco International, WorldCom, and many other large corporations, Congress imposed enhanced financial disclosure requirements for publicly owned companies as part of the Sarbanes-Oxley Act of 2002. The financial meltdown of late 2008 and early 2009 served as but the latest reminder of the ongoing need for full and accurate disclosure of such information if financial markets are to operate effectively—as well as of the need for sustained governmental oversight and regulation of these markets.

Similarly, long-standing food labeling requirements, such as calorie counts and fat and protein content, give consumers at least some of the information they need to make smarter choices about their food purchases. Estimates of new vehicle fuel efficiency, prominently displayed on rear windows, have long given automobile buyers a good idea of what to expect in fuel consumption in city and highway travel; buyers eagerly sought out that information when gasoline prices escalated rapidly in 2008. Comparable energy efficiency labels on household appliances such as washers, dryers, refrigerators, and water heaters provide similar information.

From drug safety product labels and packaging inserts to community drinking water quality reports (required by the 1996 amendments to the federal Safe Drinking Water Act) and notices about pesticide residues in food (required by the federal Food Quality Protection Act of 1996), the public's appetite for such information continues unabated. Indeed, it is extending into new territory. Increasingly people want to know about the quality of care they can expect from hospitals and physicians, the training and reliability of other professionals, and the quality of public schools and universities. Consistent with these trends, many organizations have issued "report cards" on performance in an effort to respond to the public's desire to know more (Gormley and Weimer 1999).

Demand is also rising for information about corporate and institutional carbon footprints as the nation and world begin to take climate change seriously. In 2007, the Carbon Disclosure Project, a small non-profit organization based in London, was ranking companies on their carbon emissions, and another group, Climate Counts, sought to provide similar information to consumers about how fully companies disclose their carbon footprints (Deutsch 2007). Recent reports about corporate responses to such voluntary carbon disclosures suggest that companies are persuaded to alter their energy use and set new environmental performance targets well before governments choose to intervene with regulatory requirements (Kaufman 2009b). In addition, by late 2007 a coalition of state treasurers, pension fund leaders, environmental groups, and institutional investors petitioned the SEC to demand new regulations regarding company reporting of financial risks associated with release of greenhouse gases. The coalition argued that the information was vital to investors and should be disclosed under current laws; it has not been common practice to do so.[7] This kind of information will soon be far more visible in light of the EPA announcement in September 2009 that it would begin requiring the nation's largest emitters of greenhouse gases (about 10,000 industrial sites and suppliers of fossil fuels) to track their emissions and report them to the federal government starting on January 1, 2010 (Kaufman 2009a).[8]

These varied public expectations and government mandates have a great deal in common. In a series of papers and several books, the Transparency Policy Project at Harvard's Kennedy School of Government analyzed government mandated actions that are designed to provide the public with information "to improve public health and safety, reduce risks to investors, minimize corruption, and improve public services."[9] In addition to many of the examples cited above, the project team noted the importance of international systems that track infectious disease reporting, labeling of genetically modified foods, and international financial reporting (Graham 2002; Fung, Graham, and Weil 2007). These scholars find that transparency systems have comparable components and dynamics and that their success depends on similar factors. They also find that such systems are difficult to design and maintain over time, particularly as economic markets change and information that is disclosed may become difficult to interpret.

Most of these information disclosure policies emerge from a similar normative argument that is rooted in ideas about the public's right to certain information and the government's obligation to ensure that the information is made available so that citizens can make sensible choices. Sometimes the action is taken to correct market failures, a classic example of which is the lack of sufficient information to maintain competition or to permit consumers to make appropriate choices. Requirements for information disclosure also may be seen as essential to promote equity or fairness as evident, for example, in concerns over environmental justice: the impact of environmental problems on poor and minority communities. The provision of information about toxic chemicals, hazardous wastes, or others kinds of risks can stimulate corrective action by individuals, communities, and corporations themselves. Indeed, early accounts of the TRI program tended to emphasize its potential to empower citizens and communities to bring about improvement in industrial performance through some form of public pressure on companies. The EPA itself continues to celebrate the program's effectiveness in helping to bring about sharp reductions in the release of toxic chemicals, and the program has become something of a poster child for the efficacy of environmental information disclosure requirements.

There is a less positive picture of disclosure requirements of this kind and of the TRI program itself. Even nonregulatory policies that mandate the compilation and release of such information can impose substantial costs and burdens on businesses. It is often difficult to calculate or estimate certain values, to compile the information, and to report it in the form that is required by government agencies. Similarly, despite the best of intentions, the information may not be easily understood by those it is designed to reach. Thus they may not be able to use it as intended (Gormley and Weimer 1999; Hadden 1986 and 1991; Herb, Helms, and Jensen 2003). As we will see later, a major limitation of the TRI program throughout most of its existence has been the metric on which it has relied—the amount (in pounds) of toxic chemicals released to the environment. The quantity of a chemical released is at best only a rough indicator of its risk to public health. It is a surrogate measure of what most people really want to know: how does this chemical or this facility's releases affect my health and do I need to take some action to lower the risk?

For these and many other reasons, information disclosure policies may fall short of their promise. Nonetheless, such policies are an intriguing, potentially effective, and relatively efficient way to manage some kinds of environmental and health risks. Hence they merit the attention and consideration of scholars and policymakers as one of a variety of alternatives to conventional regulation.

The Federal TRI Program: Origins and Impacts

The federal TRI program predates many of the other information disclosure policies noted above, but its origins are rooted in the same kinds of concerns about the public's right of access to critical information. The program can be traced most directly to a catastrophic industrial accident in Bhopal, India in December 1984. An American owned Union Carbide pesticide manufacturing plant there suffered a massive leak of methyl isocyanate, a highly toxic and irritating chemical, which exposed thousands of people in nearby neighborhoods. As a result, nearly 3,000 people were killed outright and at least a hundred thousand more suffered disabling injuries; many assessments of Bhopal put the death toll within one month at over 15,000. It is widely described as the worst industrial accident in history, with more than 500,000 people affected to some extent by the gas leak. Decades after the accident, its effects are still evident. Hundreds of tons of hazardous wastes remain at the site, pesticide residues at high levels have been found in neighborhood wells, and a variety of health effects are thought to be linked to the plant's chemicals (Crabb 2004; Sengupta 2008).[10]

The Bhopal incident shocked people around the world who were stunned to learn that industrial facilities could pose such an enormous risk to nearby communities and their residents. Later they learned that less dramatic chemical releases were fairly common. Indeed, less than a year after the accident, another Union Carbide plant in Institute, West Virginia that also produced methyl isocyanate suffered a leak and gained considerable media attention. In the language of agenda setting, the Bhopal accident became a focusing event or catalyst that stimulated additional media coverage of such risks, helped to build public awareness of the threat, and moved environmental activists and policymakers to press for new legislation (Birkland 1997; Hadden 1989). In terms of John Kingdon's (1995) model of the agenda-setting process, which we discuss in chapter 3, the result was a merging of the problem, politics, and policy streams that had not quite come together on the national scene before that time.

Within three months of the Bhopal accident, bills in Congress merged the right-to-know concept with reauthorization of the Comprehensive

Environmental Response, Compensation, and Liability Act of 1980, better known as Superfund. As one participant in the process put it, "The Bhopal train was leaving the station, and we got the kind of legislation we could put on the train" (Kriz 1988, 3008). Members of Congress hinted that they were also responding to a perceived reluctance of both the U.S. EPA and the Occupational Safety and Health Administration to regulate chemical hazards sufficiently, and to the limited capacity of these agencies to do much in light of substantial budget cuts they suffered in the early 1980s. The political climate at the time, particularly congressional frustration with the Reagan administration's unenthusiastic support of environmental regulation (Cohen 1984; Kraft 2010; Vig and Kraft 1984), led both the House and Senate in October 1986 to approve the final legislation, the Superfund Amendments and Reauthorization Act (SARA) by overwhelming margins.

The revised law included a new Title III, the Emergency Planning and Community Right to Know Act (EPCRA), which created the Toxics Release Inventory program. By 2009, the TRI program mandated that thousands of industrial facilities provide detailed information on nearly 650 toxic chemicals they release to the environment or transfer on or off site. In the Pollution Prevention Act of 1990, Congress further required that additional data on waste management and source reduction actions by industry be reported under the TRI program as well. The EPA also has expanded coverage beyond the initial manufacturing industries, and most recently added requirements to report on releases of persistent, bioaccumulative, and toxic (PBT) chemicals.

During the debate over the 1986 act, critics argued that Title III was unnecessary because Bhopal-like accidents were exceedingly unlikely in the United States. Yet just a few years later, a 1989 report to the EPA found seventeen Bhopal-like disasters in the nation over the previous 25 years, that is, where there was a release of deadly chemicals in volume and at levels of toxicity equal to or exceeding those in the Bhopal accident. The report tallied 11,048 accidents between 1982 and 1988 involving toxic chemicals, resulting in 11,341 injuries and 309 deaths. That the toll was not higher, the report said, was attributable to either good management or sheer good luck (Shebecoff 1989).

Even before Bhopal and congressional action in 1986, similar right-to-know laws began appearing at the state and local level as a result of many other factors, which eventually also affected the national policy agenda and subsequent legislative developments. Chief among these were continued growth in scientific knowledge of chemical and other risks

(Covello and Mumpower 1986; Hadden 1989), increasing affluence and education among the public that fostered new attitudes toward acceptable risk and a desire for greater emphasis on safety (Slovic 1987; Wildavsky 1988), and the surging memberships, resources, and effectiveness of environmental and consumer groups. These groups were now better able to mobilize a concerned public and lobby policymakers than had been the case in earlier years (Berry 1997; Bosso 2005). Perhaps equally important was a widely shared belief during the 1970s and 1980s that businesses, and particularly manufacturing facilities, should be held responsible for any harm they inflicted on the public, especially where the risks to public health were unknown to those exposed, not readily observable, had delayed effects, or were potentially substantial (Bardach and Kagan 1982; Fiorino 2006; Lowrance 1976).

The push for right-to-know laws began in the 1970s, and by 1980, Connecticut, New York, Michigan, Maine, and California had enacted laws giving workers and sometimes communities access to information about chemicals used at local manufacturing facilities. In 1981, Philadelphia adopted a right-to-know law, and several cities in California followed, as did Cincinnati in 1982. As of 1984, seventeen states and sixteen municipalities had such laws, and by mid-1985, twenty-eight states had them (Hadden 1989; Kriz 1988). By 1985, the focus shifted to the federal government in part because, as noted, Congress was considering reauthorization and broadening of the Superfund program. Industry looked with favor on such federal action because it hoped to preempt the growing number of state and local laws with a consistent national policy. Yet EPCRA specifically does not preempt state and local governments from requiring additional information from manufacturing facilities, and many do so.

Despite the federal initiatives, the states continued to approve right-to-know legislation, reflecting strong citizen concern and a belief that the states could act independently of any federal programs. Perhaps the most notable action took place in California, where in November 1986, only one month after Congress enacted EPCRA, voters approved a ballot initiative, Proposition 65, the Safe Drinking Water and Toxic Enforcement Act of 1986. It requires that citizens be informed when there is a reasonable risk of exposure to chemicals classified by the state as toxic. The act's popularity was evident in the margin of approval, 63 to 37 percent, despite intense opposition mounted by the initiative's opponents and a spending ratio by opponents over proponents of six to one.

In light of these political, economic, and social changes during the 1970s and 1980s, it is not surprising that a "risk-free" environment came to be seen as a moral issue as well, and that environmental, health, and consumer groups continued to emphasize a right to clean air and clean water, a safe working environment, and safe food and consumer products. As many students of regulatory policy have observed, the new social regulation of the era reflected a deep distrust of the business community, a desire to open the administrative process to public scrutiny, and a determination to foster increased public participation and transparency in rulemaking. A belief in the right to know about environmental pollution and other hazards emerged as part of this broader set of changes in public expectation for business and governmental decision making (Eisner, Worsham, and Ringquist 2006; Hamilton 2005; Harris and Milkis 1996).

Many policy actors were explicit in describing such a right and acknowledging the implications for the power of citizens to protect themselves. For example, former Representative James Florio of New Jersey, an author of EPCRA, said at a Senate oversight hearing in 1988 that "community right-to-know provisions will give us vital information on what it is that is out there. . . . If knowledge is power . . . those three little words—that is, 'right-to-know'—are going to be extremely powerful" (Kriz 1988, 3007–3008). Industry representative spoke in similar terms. Randal Schumacher of the Chemical Manufacturers Association (later renamed the American Chemistry Council), for example, said "I think the law [SARA] has fangs . . . . Information in the hands of the democratic society is very, very powerful." The federal law, he noted, gave the people "authority and power to change" society (Kriz, 1988, 3008).

Responding to public concern over chemical safety, in 1988 the U.S. chemical industry itself sought to improve its image and its capacity for safe manufacturing practices through adoption of a new Responsible Care initiative, borrowing elements of a similar program already operating in Canada. Over time the Responsible Care program was strengthened and eventually integrated with the environmental management systems used by companies that have adopted the International Organization of Standardization (ISO) 14001 series guidelines for environmental performance. As we will discuss later in the book, for the chemical industry the effects of the Responsible Care program and those of the TRI are somewhat hard to disentangle, but the relationship speaks to what we presume to be an important synergy of mandatory information disclosure and corporate social responsibility initiatives.

After more than two decades, what can we say about the TRI program's successes and impacts? What does its track record imply for other kinds of information disclosure programs? And what does this history say about the broader debate under way about viable alternatives to command-and-control regulation and how best to stimulate improved environmental performance?

Information Disclosure: What Do We Know?

The annual TRI reports from 1988 to the present paint a picture of substantially improved environmental performance by American industry taken as a whole. In early 2009, the EPA released data for the 2007 calendar year, and reported that for the period 1988 to 2007, total on- and off-site disposal or release of TRI chemicals decreased by 61 percent or 1.83 billion pounds.[11] These kinds of comparisons of necessity take into account only changes in the so-called core chemicals in the original industries covered by the program that have been reported on over the entire period, so they do not provide as comprehensive a measure of performance improvement as desired. Nonetheless, a 61 percent reduction in the core chemicals is impressive, as is the EPA's report on changes between 2001 and 2007. For this six-year period, the data indicate that total on- and off-site disposal or other releases of TRI chemicals decreased by 27 percent (or about 1.55 billion pounds). So the reduction in disposal or release of TRI chemicals continues, although at a lower rate than what prevailed in the early years of the program. To put these changes in chemical releases into perspective, the U.S. economy grew substantially from 1988 to 2007 (by over 95 percent in real terms), so reductions of this magnitude in release of toxic chemicals are all the more striking.

The TRI reports also include accounting for total production-related waste (TPRW). This measure refers to the total of all waste generated at a facility, or the sum of waste that is recycled on- or off-site, recovered through energy production on- or off-site, treated on- or off-site, and disposed of or otherwise released on- or off-site. For 2007, the TPWR reported under the TRI program was 24.2 billion pounds, of which 37 percent was recycled on- and off-site, 34 percent was treated on- and off-site, and 12 percent was burned for energy recovery on- and off-site. Some 18 percent (or 4.4 billion pounds) was disposed of or otherwise released on- or off-site. The amounts have changed only slightly since 2001 (U.S. EPA 2009).

Reduction in some of these toxic chemical releases is also mandated by EPA regulatory programs, such as the Clean Air Act Amendments of

1990, which included new requirements for reducing the health risks from toxic air pollutants.[12] Moreover, by one count in 2006, at least 225 industrial facilities in the United States chose to reduce their use of hazardous chemicals largely in reaction to the terrorist attacks of September 2001; environmentalists applauded the move and cited it as further evidence that such reduction in use of toxic chemicals was possible (Lipton 2006b).[13] The Bush administration also endorsed legislation before Congress in 2006 that could achieve much more. It would impose some requirements on companies to develop security plans and standards designed to limit the risk posed by possible terrorist attacks on chemical plants and other industrial facilities. According to the Department of Homeland Security, a terrorist attack on a chlorine tank, for example, could lead to more than 17,000 deaths, perhaps 10,000 injuries, and 100,000 hospitalizations. As a result of such legislation, some companies would likely consider switching to less hazardous chemicals (Lipton 2005, 2006a).[14]

The impressive reductions in toxic chemical releases shown in the TRI reports in particular help to explain why so many observers, from the EPA itself to industry groups, environmentalists, and policy scholars, have celebrated the TRI program's success. The impact of this information disclosure program is indeed remarkable, putting aside for now questions of causality. At the same time, the annual TRI reports also tell us that industries continue to release very large quantities of toxic chemicals to the environment—about 4.1 billion pounds a year from nearly 22,000 facilities across the nation; about 1.3 billion pounds of the chemicals are released to the air. Hence success measured by reduction in quantities of chemicals released over time is not altogether comforting even if it suggests the power of information disclosure to bring about meaningful change in corporate environmental performance.

It should be said that for a new category now covered by TRI reports, the EPA says that the disposal or other releases of persistent, bioaccumulative, and toxic (PBT) chemicals also remains substantial. In 2007, facilities disposed of or released some 496 million pounds of lead and lead compounds (which accounts for about 98 percent of chemicals in the PBT category), 6.9 million pounds of mercury and mercury compounds, 1.4 million pounds of polycyclic aromatic compounds, 2.1 million pounds of polychlorinated biphenyls (PCBs), and 144,729 grams (about 319 pounds) of dioxin and dioxin-like compounds. Similarly, some 835 million pounds of TRI chemicals that are known or suspected carcinogens were disposed or released, most to land disposal (U.S. EPA 2009).

If the numbers summarized above suggest that many, if not most, U.S. industrial facilities are getting cleaner all the time, there is another, somewhat less rosy, picture to paint that is consistent with the overall high level of continuing releases of toxic chemicals to the environment. A BP-owned Texas City, Texas refinery, where 15 workers were killed in an explosion in 2005, reported that it released three times the amount of toxic chemicals, including ammonia and formaldehyde, into the air in 2004 than it did the previous year. If correct, the *Houston Chronicle* argued that this estimate "belies industry claims that U.S. plants are growing steadily clean with each passing year."[15] And if the estimate is not correct, it raises serious questions about the reliability of the information industry reports via the TRI system.

In light of the data on continuing releases of large quantities of toxic chemicals and periodic questioning of the accuracy of the TRI data, as the example of the BP refinery illustrates, one might ask just how successful the TRI program has been. We address that question in detail in chapter 3. But we also have other questions that are worth asking even if the program merits the generous praise its supporters have offered. We think these questions have received far less attention than they deserve, and addressing them is critical to understanding both the potential and limitations of disclosure policies of this kind.

How does information disclosure actually affect the level of toxic chemical releases? That is, what are the mechanisms by which release of information about toxic chemicals brings about improved environmental performance at facilities across the country? There are several ways in which this might happen (which we explore in detail in chapter 2). It may be that the release of such information changes community and/or corporate knowledge and attitudes, and these changes in turn affect the management of toxic chemicals. Communities somehow communicate to local industry their desire to see reduced exposure to the chemicals, and industry takes these concerns seriously. Or, by compiling the data, industry learns something new about its manufacturing processes and changes them to improve its environmental performance. Media coverage of the reports also may make a difference; early TRI reports were often covered extensively by the local press even if coverage declined considerably in later years.

These kinds of responses also might vary from one industry to another and from one community or state to another, depending on available technologies, ease of changing production processes, the state or local economic and political environment, and community pressure. It is evident, for example, that not all facilities or all communities saw the extent of change in toxic releases captured in the summaries of annual TRI reports. So there may well be different kinds of explanations for the changes in TRI releases over time, some of which fit some industries and some communities and states but not others.

Government agencies, industries, environmental organizations, and community organizations have all made use of TRI data in many different ways to shed light on corporate environmental performance and to track community exposure to toxic chemicals (U.S. EPA 2003). Scholars have taken a keen interest in the TRI program as well, and have probed its origin, history, administration, politics, and impacts (Atlas, 2007; Graham 2002; Hadden 1989; Hamilton 2005). As a result, we know much about TRI releases over time as well as the aggregate environmental performance of thousands of industrial facilities located across the nation, and at least some of the reasons for community and industry actions and their effects. Yet many important questions remain, both about the TRI program itself and the use of information disclosure as a policy tool.

We focus on some of these relatively neglected questions. We want to know what effects the TRI program has had on communities and on the corporate facilities themselves. That is, what have been the consequences of adopting and implementing the program? For example, what difference has it made for communities that are exposed to toxic chemical releases? Are citizens better informed? Do they have a sense of empowerment? Do they communicate their concerns to local industries? If so, how have industries responded to their expressions of concern? Perhaps most important of all, are communities now exposed to fewer toxic chemicals and to the risks associated with them than was the case ten to twenty years ago?

Similarly, what difference has the TRI program made for the facilities that have to manage toxic chemicals? What transaction costs has the program imposed on business, such as the time needed to compile and report the information? What have corporations gained from the experience, such as new knowledge of their manufacturing processes and a capacity to reduce pollution releases, or the creation of better environmental management systems? What do they hear from the community in response to the release of information, say from individual citizens or from environmental or community organizations? What do they hear from the press, or from local or state government agencies? With the substantial variation from one community to another and from one industrial facility to another, we also want to know what factors most influence a facility's management of its toxic chemicals, and especially what accounts for the differences between environmental leaders and laggards. In particular, why do some corporations do so well in reducing their toxic releases and the risks associated with them while others show few signs of progress?

We believe the answers to such questions are important for the TRI program itself and to any possible policy redesign to make it more effective in informing the public, efficient in its operation, and equitable to communities and corporations. We also believe the answers speak to the broader question of the potential for using information disclosure to achieve environmental protection and other social goals, such as community health and well-being, environmental justice, and sustainable economic development. As noted earlier, we want to know about the extent to which such policy tools can supplement conventional regulation and foster not just compliance, but performance that goes beyond compliance. If the potential is real and substantial, how might information disclosure policies be designed to ensure effective implementation by government agencies? To keep the burdens and costs imposed on industry to an acceptable level? To provide the most useful information to the public?

In chapter 2, we explore the theoretical underpinning of information disclosure policies and we offer two analytical models that seek to improve understanding of how the TRI and similar program actually work. One of them focuses on the mediating factors that affect responses to the disclosure of TRI data, both within industrial facilities and within communities, such as a community's capacity to use information that is disclosed. The other, drawn from game theory, portrays the environmental performance dilemma that facilities face as they take into account the transaction costs associated with improving their management of toxic chemicals, especially going beyond compliance with environmental laws. In this chapter we also set out our major research questions, and discuss the mix of qualitative and quantitative methods we used to gather and analyze the data. In chapter 3, we focus on the history, legal requirements, implementation, and overall impacts of the TRI program. In particular, we report on the quantitative analysis of our data, which focuses on changes over time in releases of TRI chemicals and the risks associated with them. We offer several different ways to measure environmental performance, and we introduce and explain the key dependent variables that we use in subsequent analysis. One of our key findings is that although facilities have on average reduced releases and risks, there is substantial variation across the nation, from one facility to another and across the fifty states.

The following chapters turn to a more refined assessment of the impacts that the TRI program has had to date. In chapter 4, we emphasize the variability in our measures of environmental performance across the 50 states and the thousands of companies that report through the TRI program. States provide a comparative setting to examine how information disclosure and environmental performance are influenced by political and policy variability. We find that companies (and the states in which they are located) range widely in their performance over time. Some merit the "green" label while others are clearly "brown" or showing little or no improvement in performance. Among the most significant variables explaining the differences are state political conditions (such as having a strong environmental group membership), robust environmental regulations, and innovative pollution prevention policies. All help to stimulate stronger facility environmental performance.

In chapters 5 and 6, we search for explanations for why some companies and facilities are leaders and others are laggards, a question of great interest in environmental protection policy. The data we have available permit both a quantitative and qualitative review of the effects of the TRI program. Here we report on the qualitative data from our surveys, interviews, and illustrative case studies that help to explain how information disclosure actually works to bring about changes in corporate environmental behavior and in community decision making. Chapter 6 builds on this analysis by examining the distinguishing characteristics of corporate leaders and laggards. Finally, in chapter 7, we discuss the policy implications of the findings and offer a number of recommendations that we believe can strengthen the TRI program as well as comparable information disclosure policies. These are particularly appropriate in light of efforts made in recent years to address concerns raised by the business community that reporting requirements are unreasonably burdensome and costly.[16]

As these chapters make clear, we find that the TRI program and its effects are much more complex than imagined or typically described in news accounts and previous policy assessments. Release of information by no means necessarily creates an informed citizenry or a more capable community; indeed, we find that most facilities report hearing very little from citizens or community groups concerned about toxic chemicals. We also find a highly diversified set of actions by corporate America in managing its toxic chemicals. Many companies have made real progress in managing these chemicals while many others have not. Similarly, a cluster of states seems to be able to foster a higher than average level of environmental performance among the facilities located within them. Our survey data as well as analysis of the TRI data themselves speak to why these variations occur and the factors that account for the difference between corporate leaders and laggards. The findings, we believe, have significant implications for the TRI program and help to address the broader questions set out in this chapter about the potential for information disclosure and the search for a new generation of environmental policy.

2
How Does Information Disclosure Work?

In the earliest years of the TRI program, as information about chemical releases became public, many of the nation's leading corporations prominently proclaimed their dedication to reducing those releases substantially. Some of the largest companies confessed that they had no idea prior to compiling the TRI data that they were releasing such massive quantities of dangerous chemicals. The public availability of the information appeared to move them to announce significant changes in their operations and in their chemical releases. But why did they do that, and what are the implications for environmental information disclosure as a policy strategy? Among the reasons most often cited for corporate reactions to the initial TRI reports are the negative publicity the companies now faced over their newly visible chemical releases, community pressure from affected residents located near their facilities, pressure from organized environmental and other community groups—or local officials, concern from newly created local emergency planning organizations, and threats of additional regulation from state regulatory officials.

As we reported in chapter 1, TRI releases have indeed declined substantially since the first inventory of chemicals was published in 1988, even if considerable variation exists in the extent of this decline across the universe of companies and facilities. The overall downward trend would seem to legitimize the TRI program's reliance on information disclosure as a policy strategy and to speak to the future of such a strategy for addressing other environmental and social challenges, including mitigation of climate change. Can information disclosure assist the public, government officials, and even corporations themselves in achieving such societal goals? Can it bring about enough change that regulatory requirements might be relaxed, at least for some corporations that demonstrate improved performance? Might information disclosure even replace command-and-control regulation in some instances? To speak confidently about this potential, we need to know more about how the compilation, disclosure, and impact of such information works in practice.

The intriguing question of how environmental information affects decision making has long attracted interest, and for good reason. There is a widely shared assumption that such information should matter and in particular that environmental management decisions, whether in the private or public sector, ought to be based on scientific data, including careful and objective appraisals of health risks, especially those that rely on public data releases such as the TRI. That this outcome is not uniformly found has moved some scholars to ask about just how technical environmental information informs policymaking and administrative processes and what factors make a difference in strengthening that relationship (Ascher, Steelman, and Healy 2010; Keller 2009; Kraft 1998; Powell 1999; Sabatier 1978). Similar questions can be asked about corporate and community use of such information when it is distributed via a program such as the TRI.

As Laurence Lynn and colleagues (1978), among others, observed long ago, the connection between knowledge and policy decisions is tenuous and uncertain, in part because of the differing and sometimes conflicting perceptions and expectations between knowledge producers (for example, scientists, policy analysts, and planners) and knowledge users, especially policymakers, but also the public and manufacturing facilities. In a nutshell, despite a strong belief among scientists and many academics that knowledge *should* drive decision-making processes, the reality is that often it does not, nor does it necessarily even change many minds (Hadden 1991; Lindblom 1980; Lindblom and Cohen 1979).

These arguments ring true today as well, as any number of recent disputes over the use or misuse of science in environmental policy decisions, from climate change to the dangers of toxic chemicals such as mercury and lead, attest (Vig and Kraft 2010). In part because of the kinds of questions now raised about public perception of environmental risk and both public and policymaker understanding of environmental science, interest has grown in the subject of how environmental information is generated, transmitted, and used throughout the policy process (Ascher, Steelman, and Healy 2010; Herb, Helms, and Jensen 2003). We borrow from this rich scholarly tradition to offer a more focused and empirical analysis of how one kind of information, about toxic chemical releases and exposure, makes a difference within the corporate setting, communities, and regulatory agencies.

In this chapter we discuss the study's overall objectives, the methods we use, and the key research questions. We also explore the theoretical underpinning of information disclosure policies, and we offer two inter-related analytical frameworks that set out the factors that we believe affect how the TRI and similar information disclosure programs actually work. These frameworks assist us in formulating the questions that we address in the chapters that follow. At heart, we want to understand the mechanisms by which information disclosure affects decision making within the corporate setting, communities where facilities are located, and regulatory agencies, and thus over time brings about meaningful change in corporate environmental performance that eventually reduces risk to public health and the environment.

Study Objectives and Methods

The research questions that drive the study focus on the impact that the TRI program has had on corporations and communities, and particularly the former: How do facility managers respond to the TRI requirements for information collection and dissemination, and how do the facilities and the surrounding communities (that is, states, community groups, activists, and the media) use the information that is disclosed? Insofar as we can determine, what factors mediate the use of the information (such as its clarity, timeliness, utility for individuals and regulators, and accessibility), and thus condition behavioral changes and the environmental outcomes they might produce? Why are some companies, states, and communities successful in reducing the release of toxic chemicals and their risks (leaders) while others (laggards) are not? What difference do the states, the surrounding communities, and media coverage make, given that the TRI historically has emphasized the role of local communities and the right of citizens to know about toxic chemicals? What difference do the state regulatory, political, economic, and cultural environments make in corporate behavior and policy action? What do the facility managers and community leaders see as the strengths and weaknesses of the program, and what might be expected if the present program continues either as it is or with some modifications that address its current limitations?

To address such questions, we emphasize the necessity of drawing from multiple disciplines and using a diversity of complementary methods to help understand firm and facility behavior as well as state and community actions. We try to integrate the insights from theory to suggest the factors that are likely to influence both corporate and community decision making on the management of chemical risks. We also acknowledge and discuss normative questions linked to the TRI program, such as environmental justice and community rights with regard to chemical management.

Our research questions and the analytic frameworks that we present later in the chapter reflect a substantial scholarly literature that has examined variables that conceivably affect corporate environmental performance. These include facility characteristics (e.g., size, profitability, environmental expertise, experience with the TRI program, commitment to pollution reduction, and adoption of an environmental management system), community capacity to use environmental information, the level of media coverage, the stringency of state environmental regulation, and the quality of the facility's working relationship with federal and state regulators. Contextual factors also likely matter. These include socioeconomic conditions, civic capabilities and participation, and subnational governmental capacity (Delmas and Toffel 2004; Fung and O'Rourke. 2000; Graham and Miller 2001; Grant 1997; Grant and Jones 2004; Gunningham, Kagan, and Thornton 2003; Harrison and Antweiler 2003; Konar and Cohen 1997; Lynn and Kartez 1997; Prakash 2000, 2003; Press 2007; Santos, Covello, and McCallum 1996; Stephan 2002; Yu et al. 1998).

To try to answer our research questions, we draw from a diverse collection of quantitative and qualitative data. We use a large sample of TRI facilities nationwide whose releases as reported in the TRI database were tracked over a ten-year period—1991 to 2000, and an EPA computer model that translates a subset of those releases (air emissions) into a model of risk exposure for residents of the communities in which the facilities are located. That is, we try to document not only TRI releases by a representative sample of facilities over time but the extent to which those releases have either increased or decreased public health risks. To supplement these data, in 2005 we administered a nationwide survey (available to respondents in both paper form and via the project's Web page) to a smaller but still substantial sample of TRI facilities to provide an understanding of how facility managers think about the TRI program, the effects it has had on their operations, and their interaction with state officials, community groups, and the media, among others. We achieved a very respectable return rate for the survey, allowing us to use the data with some confidence in its representation of the full TRI sample. We also administered a comparable survey of federal, state, and local officials whose responsibilities included work with the TRI program and local emergency management activities to learn about their experience with and perceptions of the program; for this survey the return rate was much higher.[1]

It should be said that many critics have questioned the reliability of TRI data (e.g., Atlas 2007; de Marchi and Hamilton 2006; Natan and Miller 1998). Yet we believe that the way in which we use the information here does not present the kinds of problems that other analyses face because of the database's limitations. We use data from 1991, 1995, and 2000 (and some from later years), and we evaluate whether selection of a given time period affects the results by looking at a subsample of facilities across a larger set of years. We highlight both the conventional TRI data and the EPA's Risk Screening and Environmental Indicators (RSEI) model that the agency uses to translate the gross TRI data into more meaningful measures of human health risk.[2]

At the end of this chapter we set out our research strategy and methodology in more detail to facilitate the reader's review of the information presented in the remaining chapters. We proceed in stages from quantitative analysis of aggregate national data over time that pertains to the performance of manufacturing facilities across the 50 states to qualitative data from questionnaires and interviews—including illustrative case studies of individual facilities at selected locales around the nation. The advantages of this multifaceted research strategy are considerable.

The quantitative data are used to capture the general patterns because the number of cases is large and we can test for the range of variables that emerge from the literature and the models as noted above; the large sample associated with this approach helps to build confidence in the results. Quantitative analysis alone, however, cannot answer all of the pertinent questions. Indeed, it cannot realistically address the most important questions about why facility managers do what they do, or what they hear from surrounding communities. Thus we supplement the quantitative analysis with data from the questionnaires, statements offered by respondents as part of the survey, interviews with selected facility managers, and other information from illustrative case studies. Taken together, the qualitative data allow us to examine some of the more elusive variables and to understand better how they interrelate in the real world context of facilities, the communities in which they are located, and the actions of federal and state regulators. Our experience suggests that future research could draw profitably from both in-depth case studies that go beyond what we were able to undertake and from quantitative analysis that captures different time periods and a broader range of variables than we report on here.

As we will indicate in chapters 5 and 6, we also think this kind of study can go beyond the conventional tendency to emphasize averages and speak to the extremes of behavior. Social science research that draws from large databases typically reports the findings for the aggregate number of cases. Averages are important, and they do allow for understanding of the dominant patterns. However, the extreme cases are of interest as well, perhaps of even greater interest (Gill and Meier 2000; Meier and Gill 2000; Meier and Keiser 1996). Knowing how the average manufacturing facilities are performing and what their experience has been with the TRI program is useful. So too is learning about how the best and worst of the facilities are doing. Likewise, some policies or political conditions matter more than others in the drive toward industrial greening. These distinctions do matter, and they have consequences for public policy. Ultimately, what policymakers and administrators need to know is not just how the average facilities are performing, but what distinguishes the leaders from the laggards, and what would motivate the leaders to continue down their green paths and the laggards to change direction and improve their performance.

How Does Environmental Information Disclosure Work?

The variation in the environmental performance of facilities reporting to the TRI program can be illustrated by a quick look at just five facilities in one county of one state in the nation. Lowndes County, located in northeastern Mississippi near the Alabama border, has five facilities that reported pollution data to the EPA between 1991 and 2000. Three of them saw their air emissions go up in that period of time, while the other two saw declines. In three of the cases, the changes in air emissions were not dramatic. Two of the facilities with increased pollution levels doubled their relatively low amount of releases, while one of the facilities with decreasing pollution levels saw only an 8 percent drop in releases. In contrast, the other two facilities were a study in extremes. At the one end was the Columbus Pulp and Paper Complex, a facility that saw seven-fold increases in air emissions over a ten-year period of time. The facility went from just over 78,000 pounds of emissions to just over 500,000 pounds. Omnova Solutions, another facility in the area, was at the opposite extreme, with close to four million pounds of pollution released in 1991 and only 540 pounds in 2000, a drop of 99.99 percent.

Such extremes can be found in counties and states throughout the nation; by no means is Lowndes County the exception. What explains such variability? Has the TRI had an effect on these facilities? Was Omnova Solutions motivated to reduce its pollution levels partly due to the public disclosure of pollution information? Does information disclosure work like this?[3]

As we suggested early in the chapter, the premise of a program such as the Toxics Release Inventory is a relatively simple one. By having one set of actors provide to another set of actors information that would otherwise remain private, a dynamic is created which can change the behavior of those disclosing the information. With information comes transparency, which potentially raises the level of accountability, which in turn motivates disclosers to act in ways they might not otherwise act. This is the fundamental idea. Yet when analyzed more carefully, the nature of how information disclosure programs should work is anything but simple.

To start with, there are several complicating concerns, both normative and empirical. One normative concern is that certain information that is held by individuals or corporations is widely thought to merit protection; that is, it deserves to be kept private and not be subject to public release. Or to say it another way, does the "right to know" exist in a given situation or is there another "right to not be known"? A good example of this right of privacy is in the area of medical records. There is an understanding in the medical profession, backed up by federal law, that patients have rights of privacy that force health care professionals to be very careful in their use and distribution of medical information. Only under very special circumstances are records released to a larger audience. In sum, is there a fundamental right to know? If so, what kinds of limitations might be placed on such a right, in recognition of the need to protect privacy as well?

Another issue relates to the nature of the information being disclosed, and is linked to longstanding concerns about risk perception and communication. Can the information that is provided be easily understood by those to whom it is disclosed? If not, the value of releasing the information might be called into question. That is, what public interest is served if the information that is released will baffle more than enlighten citizens, or lead them to unwarranted conclusions, such as seeing public health risks when none exist or not seeing important risks because they are not understood (Andrews 2006; Hadden 1991; Stern and Fineberg 1996; U.S. EPA 1990)? A related question comes to mind. Who is responsible for ensuring that the information that is provided is under-standable? The corporations that compile and release it? Government agencies that consolidate and make the information public? Nonprofit organizations that choose to make the data available to the public in different, and presumably more useful, formats? The news media that may choose to publicize the information? An example that illustrates the concern is a requirement under the Safe Drinking Water Act that local governments publish annual reports of drinking water quality in their jurisdictions. Under federal law these annual reports are not only sup-posed to be widely accessible but clearly written. Information that is not easily understood raises questions about the practicality of its disclosure. In sum, is the information that is disclosed useful information, and if not, what might be done to make it more useful?

An empirical concern about information disclosure is related to the causal chain that connects information disclosure on the one hand to behavioral changes on the other. Does information disclosure change the behavior of those disclosing it without further action by those to whom the information is disclosed? Or does information disclosure first change the behavior of those given the information, which in turn changes the behavior of those disclosing it? For example, do the information disclo-sure aspects of the Food Quality Protection Act spur consumers to change their food purchasing behavior or do producers make changes in anticipation that consumers would demand them once they have the information? Potentially either or both possibilities could occur. In sum, judging how information disclosure affects behavior or decision making requires empirical study to confirm the causal relationships.

Another empirical concern relates to the way in which information is mediated as it is transferred from the discloser to other actors. Any student of communication or social psychology would readily appreciate that making information available does not mean it will directly affect individual beliefs, attitudes, and behavior, and certainly not of an entire population. Rather people must perceive and interpret the information. What they make of it and whether or how they use it will depend on various personal and social and political mediating factors (Douglas and Wildavsky 1982; Fischhoff et al. 1981; Hadden 1986, 1991; National Research Council 1989; Slovic 1987, 1993; Stern and Fineberg 1996; Wildavsky 1988, 1995). For example, information that is available to all may be read and used by only a few who find it of sufficient personal interest, say committed environmental activists or public health officials in a community. If the few who use the information are systematically different from the rest of society in important ways, this may influence the nature of behavioral changes (either the actions of the disclosers or those to whom the information is disclosed). Those who pay close attention to the information may draw certain conclusions about public health risks that are not widely shared by others. Beyond individual characteristics, such as the personal saliency of environmental risk information, other community-level or state-level variables may shape whether information is released, how it is released, how visible it is, and the impact it has on a community or state, or on the corporation releasing it. Hence we believe that a thorough accounting of the effects of information disclosure requires at least some assessment of the role of mediating factors.

When applied to the TRI program, the basic premise and complications of any information disclosure program can be boiled down to the following:

1. People have the right to know about the pollution being produced in their communities.

As we noted in chapter 1, the argument that people have such a right to know and a parallel right to participate in decisions about risk is a normative assertion long associated with information disclosure and the management of societal risks (Davies 1996; Fischhoff et al. 1981; Hadden 1989; Herb, Helms, and Jensen 2003; Lynch 1989; Shrader-Frechette 1991). The notion of a "right to know" is grounded in the belief that justice in a democratic society requires that people be made aware of the potential harms to their personal security, including their health and well-being (Stern and Fineberg 1996). Toxic pollution is by its very nature a potential harm to individuals, their families, and their communities. For citizens to be adversely affected by toxic pollution and not be able to find out why they became ill would be perceived by many as immoral or unjust. If citizens can be informed about a potential risk before harm can be done, the better off they are. This new information might help citizens to prevent the harm from occurring, in part from choosing what to do about it, including moving to a different residential area or persuading local facilities or local and state regulators to change their behavior.

2. Disclosure of information to the public creates a dynamic which spurs industrial facilities to improve their environmental performance.

The argument that the disclosure of information to the public might itself change the environmental performance of industrial facilities, regardless of other regulatory requirements, is grounded in the proposition that complete information (or the lack thereof) can have significant effects on decision making. A corollary is the argument that more information leads to better decisions, even if the information is not complete. The disclosure of TRI data is meant to give citizens a full (or at least a fuller) understanding of the toxic chemical pollution in their communities. Yet it is not only citizens who are supposed to get a fuller picture. Anyone who might benefit from such information, for example, other companies, the media, and interest groups, would also get a fuller picture, and would benefit accordingly.

3. Information is not provided in a vacuum, but rather contextual variables mediate the impact of information on the behavior of corporations and communities.

The argument that contextual factors help to shape the effects of information disclosure is an empirical claim grounded in both theories of communication and the observable variation of environmental performance found across facilities (National Research Council 1989; Stephan 2002). The contextual factors operate at multiple levels: within the facilities, within the community, and within a state. Put otherwise, information disclosure does not occur in a vacuum. Rather, it works through particular individuals, groups, and institutions. These entities may have distinctive perceptions, beliefs, attitudes, and capabilities that influence how information affects the environmental performance of facilities.

Research Questions

As we indicated initially in chapter 1, to make sense of the TRI program we need to answer a broad set of questions. The questions address both detailed causal mechanisms and the wider implications to public policy of the perceptions and beliefs of those directly involved with collecting or using TRI data.

How do facilities respond to the TRI requirements for information collection and dissemination? This question is important for understanding the immediate impacts that the TRI program has on the behavior of manufacturing facilities. One way of answering the question is to analyze the political actions of those businesses being asked to relinquish the relevant information. Are they willing or reluctant participants? To what extent do they find the TRI information requirements to be a burden, and would they favor relaxed requirements or none at all for such information disclosure? Corporate efforts to persuade the EPA to limit the scope of required reporting partially captures the level of animosity the requirements have received in the business community, but the level of support or opposition cannot be judged thoroughly without empirical study. Another way of addressing the question relates to the behavioral changes that facilities (and firms) have made within their organizations in order to be able to competently and efficiently collect the relevant information. How have the management practices of facilities been changed by the need to collect TRI data? Changing guidelines, additional resource availability, and the hiring of new personnel may all indicate that facilities have moved to build their capacity to respond to TRI requirements.

How do facilities, their surrounding communities, and public officials use the information that is disclosed? This question is at the heart of what we explore in the book. Arguably there are numerous ways the information could be used, including the possibility of non-use of the data; that is, information is collected and made available and there is no use of it at all, neither internal to the facility nor within the community or the state. At the other extreme, the information is of keen interest to the facility and to the community, and it is a vital part of chemical management decisions both within the facility and within state and federal regulatory bodies. We believe the great variability in the environmental performance of facilities hints at the complex ways in which information might influence behavior at both the facility and community level. We try to clarify how information is used, by whom, and with what effects.

What factors mediate the use of information, and thus condition behavioral changes and the environmental outcomes they might produce? If knowing how TRI data are used is at the heart of the book's purpose, then understanding the role of mediating factors is the fundamental social science puzzle that drives the research even if we can address it only indirectly. The disclosure of TRI data, like any information disclosure, appears not in isolation, but instead as part of a larger narrative which shapes the understanding of all who are aware of it. Information dissemination occurs within particular communities, under various institutional settings, and at specific historical moments. Users of information bring their own experiences to bear. Some factors are likely to positively affect behavioral changes, while other factors are as likely to stifle change. Making sense of these factors helps to increase our understanding of the complexities of the TRI program and the mechanisms by which disclosure of chemical risk information affects decision making.

By pulling together the answers to these key inquiries, the following question can in turn be addressed: *why are some facilities, communities, and states successful (that is, leaders) in reducing the release of toxic chemicals and their risks while others (that is, laggards) are not?* What combination of factors seems to make the difference? As we noted earlier in the chapter, to answer these questions we rely on a research strategy rooted in both quantitative and qualitative approaches that combine aggregate data analyses with case studies and surveys. Our goal is to understand both the macro picture—the overall patterns in environmental performance over time—and the micro picture—finding reasonable explanations for variability from facility to facility, community to community, and state to state. Ultimately, by focusing on leaders and laggards across facilities and states, we hope to offer a more nuanced view than previously available of how the TRI program affects behavior and changes environmental performance.

Finally, *what do the facility managers and others see as the strengths and weaknesses of the program, and what might be expected if the present program continues as it is or with some modifications that address present limitations?* This question is important because it pertains to some clear opportunities for policy change in the TRI program that could respond both to industry concerns and to citizen interest in having a reliable and useful source of information about chemical risks present in a community. In addition to advancing knowledge about the TRI program and its effects, and the larger challenge of information disclosure as a policy strategy, we hope the findings reported here also have practical value for those interested in policy change. Since its inception in the 1980s, the TRI program has undergone many changes. The list of chemicals whose use and release must be reported has grown substantially (from 270 chemicals in 1988 to nearly 650 chemicals in 2009), and administrative rules and procedures related to how information is collected and reported have changed as well, at both the federal and state levels. Yet as we discussed in chapter 1, the volume of toxic chemicals that continues to be released remains very large, with potentially severe public health risks. In chapter 7 we will review the policy implications of the research and offer some suggestions for how the TRI program might be made more effective, efficient, and equitable.

We begin the process of addressing these research questions by investigating what theory tells us we should expect and what previous research has already discovered. The theoretical constructs and research results of relevance come from a variety of sources. No single discipline, nor theoretical argument, serves to address all of the important questions raised above. Rather, multiple disciplines must be tapped in order to fully make sense of the TRI and of comparable information disclosure programs.

The Right to Know: History, Theory, and Research

Although contemporary use of the concept "right to know" dates primarily from the 1970s and 1980s with the growth in federal and state consumer and environmental protection policies (Eisner, Worsham, and Ringquist 2006; Harris and Milkis 1996; Lynch 1989), its history is much older. Indeed, in some respects one can trace the concept to the nation's founding and the idea that average citizens have a right to know what their elected leaders are doing on their behalf. Elected officials in particular are expected to be held accountable for their actions, and for that reason there is a long tradition of conducting government affairs in a transparent manner that permits citizens reasonably full knowledge of decision making. At the same time that the nation's major environmental policies were set in place in the early to mid-1970s, public expectations grew for open and accountable government, evident, for example, in campaign finance legislation, open committee hearings in Congress, and advancement in making information about the operation of government agencies and policymaking bodies available to the press and the public (Gormley and Balla 2008; Kraft 2011; Williams and Matheny 1995).

As we recounted in chapter 1, in recent decades the idea of the right to know has expanded beyond elected representatives and now entails any information that a particular organization or sector of society might have a moral responsibility to share. The notion of accountability has widened to include the accountability that different community members or groups have to each other. As noted earlier, the right to know is in creative tension with the right to privacy. Not all information is required to be transparent. Any policy or law that delineates what should be known also implicitly or explicitly draws a line to distinguish what information may remain private. Critical to the drawing of that line is the issue of harm—will someone be harmed if the information is not released? Will someone be harmed if the information *is* released? When framed this way, the importance of the right to know as it relates to the accidental or intentional release of chemicals into the environment makes perfect sense. People should have the right to know about activities that have the potential to do them physical harm.

There is little doubt that the tragedy in Bhopal, India, in December 1984, sparked the development of federal right-to-know legislation in the United States (Hadden 1989; Lynch 1989). A similar, but smaller scale incident at another Union Carbide plant in West Virginia accelerated regulatory action. These accidents and the enormity of harm inflicted on people in Bhopal were shocking to many. They had little awareness that manufacturing industries, both in the United States and around the world, could pose such an enormous risk to nearby communities and their residents. But the Bhopal and Institute, West Virginia accidents awakened them to that risk. As we stated in chapter 1, the accident drew enormous media coverage and catalyzed public policy action in a way that reflects scholarly commentary about the general effects of accidents or focusing events of this kind as well as John Kingdon's widely cited argument about the convergence of information about public problems, available and tested policy ideas, and political willingness to act (Birkland 1997; Kingdon 1995). One result was overwhelming congressional approval in 1986 of the Superfund Amendments and Reauthorization Act (SARA), with its new Title III, the Emergency Planning and Community Right to Know Act (EPCRA), and the new Toxics Release Inventory program.

The development of SARA and the TRI program was one of the first sustained attempts by the federal government to address environmental information needs in a programmatic fashion (Hadden 1989; Hamilton 2005). The TRI meant going beyond the platitudes and actually institutionalizing the transfer of information from one segment of the society to another. It is easy to say the public has a right to know, but what does this really mean, and how many people will want to exercise such a right, and will do so with care? And how will they be able to deal with the enormous and confusing set of data that TRI presents? EPCRA and the TRI fall to the weak side of the powers that governments may exercise (Anderson 2006; Gormley 1989; Kraft and Furlong 2010). They are intended to provide information, but not to do anything about the chemical releases that are not already governed by other environmental laws, such as the Clean Air Act or the Resource Conservation and Recovery Act. So whether such information disclosure policies can do much to protect the public's health and well-being depends heavily on what manufacturing facilities, regulatory bodies, and the public itself choose to do. As we explain below, the effects of information disclosure policies like these depend on the interaction of a number of variables, particularly the degree of interest and understanding on the part of the public. In chapter 3 we speak to the implementation and impacts of the TRI program.

In terms of the public's right to know, what developed out of TRI and related programs was a sense that the public had the right to information about the *estimated amount* of pollution being created by industrial facilities in their communities. This information could be gathered at the chemical level and was required of any facility that had ten or more full-time equivalent employees, was in a covered industry sector, and crossed any of the thresholds for manufacturing, processing, or using any of the listed chemical during the year.

Yet by focusing attention exclusively on the estimated amount of pollution, the EPA was able to avoid some potential problems. First, by using estimates, the agency was reducing the potential burden on facilities that might not have the means to carefully measure their pollution levels. Estimates, based on reasonable parameters, were deemed to be enough. Second, by focusing on the amount of pollution, the EPA was avoiding the problematic process of calculating the risk created by the industrial facilities. Lobbying by industrial interests indicates that they tried to have it both ways. On the one hand, there was a strong desire to reduce their data collection burden and they worried that too many details would be overly burdensome and costly. On the other hand, some industrial interests believed that a uniform federal law would be better than the haphazard range of requirements that began appearing at the local and state levels during the 1980s (Hadden 1989, 28). Furthermore, some corporate interests complained when the EPA first began releasing information that just releasing data about emission amounts could be seriously misleading for the average citizen, who had little or no capacity to interpret the information other than to imagine that large releases posed a risk to their health (Hamilton 2005, 55). As we will see in chapter 3, the decision to focus on pollution amounts (pounds) rather than risk may have distorted public understanding of the environmental performance of industrial facilities and diminished what could have been a stronger citizen and community role in chemical management.

Information and Behavioral Changes: Theory and Research

No theory is more central to understanding information disclosure programs than the notion of *transaction costs*. The idea, arising out of economic theory and the work of Ronald Coase (1988) in particular, is that any bargaining that occurs between two actors in the real world cannot occur without costs. The reality of costs means that sometimes bargaining will not be efficient or effective, and that sometimes it will not occur at all. What is the relevance to the TRI? The existence of transaction costs, and the resulting information asymmetries (Williamson 1985), helps to explain the need for an information disclosure program in the first place. Such programs help to reduce the costs of information acquisition by all parties interested in bargaining over a contract or negotiating a conflict. In the case of the need for TRI data, the conflict has to do with the negative effects of pollution downwind or downstream from a polluting facility. Government's provision of detailed toxic release information lowers the costs that otherwise might have been born by the community itself. These lower costs, in theory, should mean that communities might be more likely to act to stem the negative effects of pollution.

The pollution created by the facility can be understood as an *externality*, a by-product of the manufacturing process that can cause damage to others not directly involved with the facility. Citizens living near a polluting facility might be keenly aware of the pollution being created, but may not have a clear sense of the extent of environmental damage or public health risks associated with it. The provision of TRI information may help to clarify for citizens the potential harms and risks created by the pollution. The costs to facilities for providing the pollution data may serve as an informal externality charge. Without the requirement of transparency, the communities near the facilities would shoulder the costs of improving their understanding of the harms being created; with the TRI, some of the costs are internalized within facilities.

Transaction costs and the government's rationale for informational intervention fall within the broader set of collective action dilemmas, including regulation. Government serves not as a "coercer" but rather a "facilitator" (Scholz and Gray 1997). Building implicitly or explicitly on the ideas of Anthony Downs (1957), Mancur Olson (1971), and Ronald Coase (1960), the argument has been as follows: the reduction of information costs increases the likelihood of participation by all actors affected by pollution output. Downs discusses information costs in terms of voting behavior of citizens and the implicit cost–benefit analyses that they perform when considering whether to vote. Olson expands the idea of information costs to any form of collective action, and he argues that political actors will only bear the costs of information collection if they gain selective benefits for doing so. Furthermore, non-state actors are limited in their forms of leverage in ways that governmental entities are not; for instance, interest groups do not possess the full force of coercion that the state does. Finally, through explicit introduction of transaction costs, Coase argues that the efficiency of bargaining between two actors whose preferences are in conflict is undermined by the reality that bargaining itself has costs that may limit the likelihood that it will be used as well as its effectiveness and efficiency. How do these ideas relate to information disclosure programs?

Recent work helps to explicate the connection. Potoski and Prakash (2004), following Scholz (1984, 1991), described the environmental regulation version of the classic prisoner's dilemma. Theory would predict that most of the time, regulators and industry will be individually rational and collectively irrational. For instance, industrial companies have no incentives to reduce more pollution than they have to because their current information leads them to expect government to continue to emphasize compliance with existing regulations. Likewise, federal and state regulators are not likely to support industrial greening with regulatory relief because they expect industry merely to comply with regulations and not to go beyond them (Press and Mazmanian 2010). This dilemma can be exacerbated by the lack of environmental performance information. Fiorino (2006) described this "compliance imperative" as a major barrier to progress in industrial environmental governance because, as we discussed in chapter 1, it is by far the prevailing institutional mindset of command-and-control regulation. Any move toward industrial greening, according to game theory, looks suboptimal to both of the players: industry and government. Therefore, both become prisoners in a collective action, or "regulation dilemma." Information asymmetry is one of the main causes of this dilemma.

The importance of transaction costs is tied closely to the economic concept of information asymmetry. In practice an *information asymmetry*, a gap in knowledge between two or more actors (Williamson 1985), exists between a polluting facility and its surrounding community. The facility is likely to know much more about its chemical use, its production processes, and the pollution outputs associated with its manufacturing activities than anyone else who resides in the surrounding community. If information acquisition were costless or relatively so, then residents could overcome their information deficiency quite readily. Yet it is because this information acquisition is not costless that government's provision of information serves as a subsidy of some importance. The provision of information brings the two key parties into greater (if not perfect) balance. Any concerns raised by the community would be better informed with TRI data than without such data.

Government's subsidy of information acquisition becomes critical only to the extent that new information is understood to have a *motivational effect* on those who are receiving the information. In particular, when it comes to pollution information, there is reason to believe that new information might serve as a catalyst for action for community members near a TRI-reporting facility. Because of the fear or dread of the negative consequences of pollution exposure (Covello and Mumpower 1986; Douglas and Wildavsky 1982; Slovic 1987), citizens would be expected to put pressure on nearby facilities to reduce their pollution output. Either citizens would have a greater sense of risk (and would want to reduce that risk) or they would have more knowledge about the particular chemicals being emitted and would use this knowledge to make their case for reductions. Barring other constraints, psychological theory would suggest that the greater the fear of the citizens (up to a reasonable limit), the greater the likelihood of action to address the fear (Fischhoff et al. 1981; Lowrance 1976; Perrow 1999; Slovic 1987).

There are a number of possible behavioral changes that can occur with new information about toxic pollution. First, citizens can choose to act in ways that meets their needs. They might put direct pressure on facilities in their communities through a variety of means, including legal action through the courts, appeals to public opinion (for example, via letters or other communication to the media), or direct communication with the facility's managers. Second, other actors may act in place of citizens. For example, the local media may report on the toxic pollution of local facilities, informing the citizens in an easily accessible manner, or environmental activists may use the TRI data in an effort to mobilize supporters in the community.

Third, the dynamic aspect to information disclosure can arise when facilities try to anticipate the possible reactions of nearby communities to new pollution information. If a facility believes there is even a slight chance of community attention, it may be willing to move ahead with pollution prevention in order to avoid or mitigate such community scrutiny and possible criticism. The cost of doing so would be deemed a reasonable business expense in light of the possible negative reaction by the community should pollution releases become visible and a matter of local contention. The logic here is similar to the anticipatory action of elected officials who assume attention to their activities and retrospective voting on the part of their constituents (Jacobson 2009; Mayhew 1974). A retrospective voter reviews the past actions of an elected official and uses this information to decide whether to vote for or against the incumbent. Elected officials who want to be reelected would be motivated to act in a way that would please a majority of likely voters. Both industrial facilities and elected officials would rather head off controversy or displeasure before it occurs, if at all possible.

Fourth, facilities may anticipate local action and act in a way that attempts to mitigate the likelihood of citizen behavior or even the behavior of other parties. A facility might work to reduce its pollution levels as a way to counteract any negative press it would otherwise receive. We have described this anticipatory performance elsewhere (Abel, Stephan, and Kraft 2007). Of course, any or all of these possibilities could occur simultaneously. Information provision could have a ripple effect at a number of levels.

There is another possibility for how information disclosure can be dynamic. In this case, requirements by the EPA that facilities report their pollution levels would force facilities to gather data which they are not already collecting. In so doing, the EPA would cause facilities to give attention to an issue they might otherwise ignore. Here internal mechanisms might shape the institutional response, regardless of what is occurring outside of the facility. The business adage, "You manage what you measure," would apply. The internal drivers within a facility (or firm) might come from a variety of sources. Upper management might make a commitment to better environmental practices for the firm or facility. Environmentally-minded employees might use the new data as a basis for pushing the facility to change its practices. Or the evidence of significant releases might move managers and employees to seek new technologies or changes in production processes that limit the releases simply as a routine business decision that saves money for a facility by improving its efficiency of operation. Reductions in toxic emissions could occur without any outward sign of pressure, or even a sign that facilities would be expecting to receive such pressure.

The literature on the behavioral effects of the TRI program or similar information disclosure programs has come to varying and not entirely consistent conclusions. Some evidence suggests that citizens can be and have been spurred to action by the release of pollution information (Beierle 2003; Bouwes, Hassur, and Shapiro 2001; Fung and O'Rourke 2000; Herb, Helms, and Jensen 2003; Lynn and Kartez 1997). Other studies have found that although citizen action may occur in isolated incidents, the bulk of facilities that have reduced their pollution output have done so in anticipation of a negative backlash from members of their communities (Lyon and Maxwell 2004).

Some evidence from studies of capital markets and media markets (Hamilton 1995) suggests that rather than citizens acting directly in their own interests, intermediaries have done the heavy lifting when it comes to reacting to new pollution data. Both stock market dips and media exposure can be understood as signals to facilities to move toward cleaner production. These signals may serve as early warnings of further action that could occur without a proper response.

Finally, the data we present in this volume are consistent with arguments by others (Coglianese and Nash 2001; Lyon and Maxwell 2004) that many facilities have been spurred to action because of internal reactions to information gathering and dissemination. The collection of data that is mandated for facilities forces them to confront their own behavior in ways that did not occur before the introduction of information disclosure programs.

Information disclosure, as a form of public policy, also can be understood to be what Schneider and Ingram (1997) called a capacity-building tool. By informing or enlightening people, it acts as a partial step toward empowering people to act. This concept of "capacity-building" can be seen as part of the wider context of variables that would be expected to help increase the capacity of facilities to reduce toxic emissions and variables that would be expected to decrease the capacity of facilities (see also Weiss and Tschirhart 1997).

Mediating Factors: Theory and Research

The argument that contextual factors help to shape the effects of information disclosure is an empirical claim grounded in the variation of environmental performance found across facilities. An exhaustive list of all possible influences would not be very helpful if they cannot be examined empirically, but certain factors stand out and can be studied.

Capacity within Corporate Facilities
The capacity of facilities to progressively manage their pollution output is based on a number of factors. Partly it is about their expertise in environmental management. The number of staff, their training, their experience, and their access to resources, particularly networks of environmental professionals, all are likely to influence environmental practices.

This expertise may also come from experience. It can be expected that a facility that has dealt with the TRI program for a number of years would be better prepared than a facility filling out a TRI form for the first time. Particular staff experience may also matter. Those filling out TRI forms year after year will have a lead on those new to the reporting process no matter what the experience of the facility as a whole with the reporting process.

Yet expertise is only part of the equation. Another factor is the beliefs of executives and managers within facilities. The extent to which there is a corporate commitment to better environmental stewardship or perceived economic benefits from pollution reduction are both examples of how beliefs can matter. If management is highly concerned about the facility's reputation or expects new regulatory action in the near future, these beliefs could also influence the willingness of facilities to move beyond their legal requirements and reduce releases of toxic pollution in a more dramatic fashion. These beliefs can be broadly understood as forms of *corporate environmentalism or corporate social responsibility* (Esty and Winston 2006; Hoffman 2000; Press and Mazmanian 2010) to the extent that facilities have embraced better environmental practices as part of their corporate model, this can be understood as a fundamental aspect of facility capacity.

Market Reactions

Closely related to corporate capacity is what we refer to as market reactions. One component here is communication of corporate actions through the media, particularly local newspapers and television stations. Media coverage of chemical management decisions and of the TRI reports might well make a difference in what facility managers choose to do. We believe this was especially the case in the early years of the TRI program when such information was more newsworthy than it is today. Over time we would expect media coverage to be of decreasing importance as TRI reports no longer attract the attention they once did. Indeed, our corporate survey data and case studies strongly reinforce this suspicion.

Similarly, and particularly by the 2000s, financial markets may be quite sensitive to information about environmental performance in addition to the obvious importance of financial performance such as profitability and growth potential. Green companies might well be rewarded by investors (including mutual fund and pension plan managers), and they may develop better relationships with their suppliers and customers as a result, or they may be under some pressure from their customers (again, particularly in recent years) to improve performance. Conversely, markets may penalize facilities or firms with poor performance (Baue 2008; Press and Mazmanian 2010). Astute facility and firm managers would likely be sensitive to such possible reactions and engage in a form of anticipatory planning.

Of course, there may well be tension between economic and environmental performance even if contemporary emphasis on achieving sustainable development reduces that tension to some extent. As we discussed earlier, the transaction costs of compiling and reporting TRI data are not insignificant, even if in many cases those costs are lower today than in the early years of the program. There is less doubt that in the 1980s and 1990s, including much of the period we use for our data analysis, many firms and facilities would see improving environmental performance to be in some conflict with the economic bottom line. Hence, all else being equal, we would expect to see more interest in environmental improvements, such as reduced toxic chemical releases, on the part of firms or facilities that are in sound economic shape.

Capacity within Communities

Community-level and individual-level variables can help to shape what happens within facilities as well, although perhaps most importantly these variables shape the reactions of citizens and other community members to the information passed on from the TRI. Whether and how citizens use new information is likely to be compounded by social, economic, and political factors that vary from community to community and state to state (O'Rourke and Macey 2003; Press 1998, 1999). Take, for example, the capacity of a community to make sense of complex technical information such as the risk from a thousand pounds of a particular chemical released to the atmosphere by a facility in a given year. Certain communities are going to be more likely than others to be able to makes sense of such information. Capacity in this sense is likely to be a factor of the average education level of community members, access to local sources of scientific expertise and experience (such as at a college or university), prior experience in dealing with pollution matters, the capacity of local citizen and environmental groups to acquire and act on the information, and even the quality of the local media and their ability to cover stories related to such information release.

Critical to understanding community capacity are the resources within communities that can be translated into civic or political action. Socioeconomic conditions would arguably be fundamental. Levels of civic or political participation in local matters might also reflect the ability of communities to act on pollution concerns. Community leadership in the form of environmental groups, civic organizations, local officials (mayors, members of city councils, or county officials), and local emergency planning committees (LEPCs) might be particularly important in this regard. Media coverage and/or the perceived seriousness of an environmental problem could also spur citizens to act in ways they otherwise would avoid.

It is important to note that the capacity of the community for action need not be realized. It only needs to be a potential. To the extent that facilities want to stay ahead of community action—stopping it before it starts—the potential of a community to act could be quite telling. Community capacity can therefore be understood to be a "latent" capacity, one which could be activated, but need not be in order for changes to occur at facilities within the community.

Subnational Governmental Capacity
The expertise, resources, and beliefs of governmental actors, whether local or state, could also be expected to be of relevance. Capacity in these cases could be measured partly by past actions, e.g., budget prioritization on pollution prevention. It could also be understood to be a function of the general commitment made by government to policy positions in line with the logic of pollution prevention. We might expect that government commitments to certain types of policies and programs, for example, health care and social welfare, might correlate well with commitments to environmental policies and programs.

There are also *state-level* variables that help to explain why certain facilities perform better than others. State-level factors may be institutional, but they also may be connected to mass politics in the form of public opinion or political ideology. States are critical politically-bounded units that shape policy choices (Erikson, Wright, and McIver 1993; Rabe 2010; Ringquist 1993a; Scheberle 2004; Teske 2004). In the study of state regulatory politics, principal agent theory dominates the empirical work examining environmental policy. Borrowing from economics, the principal-agent model in political science characterizes a whole series of relations where principals delegate to agents. Citizens vote for politicians, legislators assign work to committees, political appointees entrust their agency staff, and regulators even delegate to the regulated. According to Gerber and Teske (2000), principal-agent theory is a dominant

Table 2.1
Performance Dilemma

	Facility	
Government	Compliance	Beyond Compliance
Encouraging (Preventing)	(Mild Booster) (4,1)	(Performance Synergy) (3,3)
Commanding (Controlling)	(Minimal Expectations) (2,2)	(Corporate Citizens) (1,4)

Note: The first number in each box represents the payoff for government and the second number represents the payoff for the facility. The payoffs are consistent with standard restrictions placed upon prisoner's dilemmas (Scholz 1991, 118).

approach to understanding how much politics influences bureaucratic behavior and the policy choices that state agencies pursue. These policy choices in turn can be understood to influence outcomes such as pollution production.

Our Analytical Frameworks

Pulling together what has been described above in terms of various frameworks and theories that have been found helpful in explaining corporate behavior and community and individual action on chemical management, we use the analytical frameworks pictured in table 2.1 and figure 2.1 to guide our work. Both frameworks are directly relevant to our research questions.

Table 2.1 presents our basic understanding of the environmental performance arena within which facilities and governments (federal and state) find themselves. The framework is a variation, with some important additions, of Scholz's "enforcement" dilemma (1991) and Potoski and Prakash's "regulation" dilemma (2004).[4] Facilities can meet their regulatory requirements for toxic emissions or they can surpass those requirements. Governments can encourage facilities to move beyond compliance in a variety of ways—through voluntary programs, enhancing community involvement, supporting interest group activities, or spotlighting performance for media attention—or governments can place their primary focus on compliance with environmental laws and administrative rules.[5]

In the absence of pressures from other actors or factors outside the dilemma, the payoff structure leads to an equilibrium where both governments and facilities will be motivated to follow the traditional

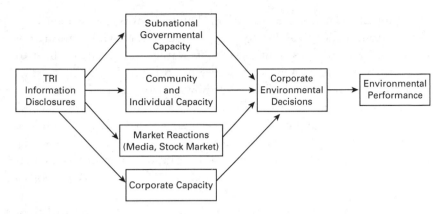

Figure 2.1
The Effect of TRI Information Disclosures on Corporate and Community Decisions

path of a command-and-control relationship. This equilibrium is not optimal; both governments and facilities are better off strengthening performance. Information disclosure programs, such as the TRI, have the potential to move facilities and governments toward performance synergy: the choice below where the greening of industry can result in the most progress.

Although helpful for analytical purposes, the table oversimplifies the outcomes that broader theory would predict.[6] Arguably, because a number of factors can influence the behavior of governments and facilities (and influence the pay off structure), it may be possible for any given facility to fall into one of the four boxes. In cases where governments focus on encouraging facilities, but facilities focus solely on compliance, the actions of governments can be seen as weakly cheering on facility behavior while facilities themselves do just enough to meet legal requirements. When both governments and facilities focus on minimal standards, performance itself does not exceed threshold expectations. Facilities that reach beyond compliance without governmental encouragement may get a pat on the back, but no other credit is forthcoming. Finally, when both governments and facilities focus on increasing performance, the rules and regulations set only a baseline to strongly surpass. We should add that, as Daniel Fiorino (2001) has argued, environmental policymakers and corporations, among others, learn from their experience. Any model of policy behavior and policy change ought to incorporate such learning, which is often rooted in "knowledge acquisition and use."[7]

Figure 2.1 clarifies the role of mediating factors that enhance or detract from the TRI's ability to motivate facility behavior. Information disclosure does not occur in a vacuum. Rather, other factors help to shape the desire and ability of facilities to go beyond compliance. TRI data are made publicly available, but how or whether that information is used is partly determined by the resource capacities of actors such as state governments, community groups, regional or local media outlets, and toxics emitting corporations. Resources can be financial, but they also can be organizational (e.g., how a corporation structures its internal decision-making) or political (e.g., the ability of community groups to interact with public officials).

Based on these analytical frameworks and our discussion earlier of the theories and ideas pertaining to corporate and community behavior, we can begin to address the research questions laid out earlier in the chapter. In particular, our general sense is that as facility, community, or government capacity increases, so too should environmental performance. Increased capacity in any of these domains may motivate changes in corporate environmental decisions, which in turn may connect to performance. In chapters 4 and 6 we analyze some of the specific factors related to capacity, including the characteristics of state government, management commitments to environmental practices, and the intensity of contact between facilities and community groups. Some of the relationships we note here can be studied only as additional data become available.

The Data

Any attempt to make use of the TRI dataset must start with the basic fact that the dataset is large and detailed. For 1988, for example, the first year in which consistent data can be easily accessed, there were over 270 chemicals reported across roughly 3,500 facilities nationwide. By 2009, there were nearly 650 chemicals reported by over 22,000 facilities. The TRI not only includes total pounds of pollution produced by chemical by facility, but it also breaks these data out based on the nature of the release (air, land, or water) and disposal (e.g., off site or on-site, injection, treated or untreated). Data can be aggregated and disaggregated in numerous ways, adding to the complexity of analysis.

Because of the size and complexity of the TRI, making sense of it for research purposes is by no means a straightforward process. What should be included and what should be left out? Our answer has been to look both as widely as possible and as narrowly as possible. We describe national trends over time, but also look at facility-level data for particular facilities in order to make sense of the pattern. Although we look at data from 1988 to 2007, our core dataset is intentionally limited to three distinct years.

First, we look at three years, 1991, 1995, and 2000, in detail. This allows us to examine trends over time in a manageable way, while at the same time allowing us to make facility comparisons within a given year. Chapter 3 lays out in greater detail our reasoning for choosing these three years.

Second, we analyze information for facilities that report in all three years. In this ten-year period of time, some facilities stopped reporting TRI data, while other firms started reporting after having not reported in previous years. Looking at just those facilities that report in all three years allows for a more consistent comparison in changes over time at both aggregated and disaggregated levels. For example, we were able to examine particular facilities over time, but also aggregate that data to the state level for further comparisons.

Third, we analyze across a consistent set of chemicals in the three years of data. Over time, the TRI program has added and subtracted chemicals from the reporting list. By examining a consistent set of chemicals, we avoid an "apples versus oranges" scenario in which facilities look worse or better in latter years because they are now either reporting information on chemicals they were not required to report on in previous years or not now reporting information on chemicals they were required to report on in previous years.

Fourth, we put less emphasis on the exact amount of pollution being produced in any given year and give more attention to the trends upward or downward over time. Some critics have suggested that because of the self-reporting nature of the data, there may be serious deficiencies in their quality (Graham and Miller 2001; Natan and Miller 1998). Although this is a distinct possibility, by giving more attention to trends than raw numbers, we have been able to avoid some of the problems that come with using potentially inaccurate data.

Fifth, we use both conventional TRI data about toxic pollution releases and data gathered from the new RSEI model described earlier, which allows a more nuanced and meaningful analysis of risks to human health. Although release data remain important, the addition of risk analysis adds an extra layer of critical information that goes beyond what was available to facilities in the 1990s.

The importance of the RSEI model to provide new and different information about pollution cannot be overstated. One of the greatest limitations of the TRI has been the release of information by pounds of chemicals, only a rough measure at best of health risks. Insofar as public understanding and corporate actions should really be based on such risks and not merely on the quantity of chemicals released, this model is a welcome addition to the set of tools available for understanding pollution effects.

Conclusion: The Logic of Our Research Strategy and Methodology

As we discussed earlier in the chapter, to make sense of the TRI program and the environmental performance of facilities over time, we incorporate a multi-method approach. Using a combination of quantitative secondary data analysis, survey data, and interview data, we proceed in stages, moving from the widest possible angle to the narrowest. Our work begins with a detailed descriptive examination of release trends over time from the very beginning of the program until its most recent results. In so doing we give a sense of the complexity of the data under study, while also giving some indication of how the performance of facilities has changed over time.

We next turn to a dataset which allows use of inferential statistics to help answer our research questions. Using multiple dependent variables and an array of independent variables, we test variations among facilities, but also across a subset of the 50 states. Such analyses allow for patterns to be discerned and for the relative weight of different causal explanations to be compared. Standard regression techniques are combined with path analysis to better understand both the direct and indirect linkages between facility pollution performance and other facility characteristics, economic factors, political factors, and social factors.

The quantitative data analysis is supplemented with data from our surveys of facility managers and public officials directly involved with the TRI program. The surveys allow us to ask questions that cannot be answered by the TRI data themselves, such as the relative importance of factors affecting chemical management, communication with community groups, media coverage, and the effects on the facility of TRI data collection. Interviews and illustrative case studies round out our dataset and serve to enhance the results of the surveys. Using questions similar to those incorporated into the survey, but more open-ended, we gathered data about the efficacy of the TRI program as perceived by both those who have to fill out TRI forms and those working within federal, state, and local government on TRI related topics. This kind of qualitative data fleshes out what cannot otherwise be fully understood.

Taken as whole, the methods are multi-layered, allowing for a sense of context not usually found in analyses of this kind. Rather than isolated segments without a coherent relationship, the process here described is one of mutual reinforcement and interrelated perspectives. For example, as we noted earlier, while quantitative techniques do a better job of looking at average cases and common patterns, the survey work and interviews allow extra attention to be paid to the extremes. In this case the extremes are the leading and lagging facilities, that, when compared, accentuate the distinctions that matter when it comes to pollution performance.

It is worth stating that despite the comprehensiveness of our review of the TRI program, there are no simple answers and no simple recommendations to make. We hope to provide useful contexts and point to constructive changes that would make the program, and others like it, more effective. Readers can judge the results for themselves in the subsequent chapters.

3

Reducing Toxic Releases and Community Risks

One of the most striking developments upon the implementation of the Superfund Amendments and Reauthorization Act (SARA) and release of the Toxics Release Inventory (TRI) was the flurry of announcements from major corporations suddenly attuned to new public expectations for coming clean, that is, for improvements in environmental performance and transparency in release of toxic chemicals. One company that received more press attention than most others was Monsanto. In the early 1990s it was a leading chemical manufacturer and today it is one of the largest and most prominent agricultural biotechnology corporations in the world. In 1988, the first year of reporting for the TRI, Monsanto officials realized that the very large number they would soon be reporting could be detrimental to the company's reputation. They chose to head off the expected negative publicity over those numbers by initiating a voluntary emissions-reduction program. The company's CEO, Richard Mahoney, publicly embraced the goal of corporate pollution reduction, and announced that his company was determined to cut its toxic air emissions by over 90 percent within five years. And it did so ahead of the target date after spending some $100 million on the effort (Dahl 1997).

These kinds of changes are by no means confined to the nation's largest and best known chemical companies such as Monsanto. Striking examples abound of comparable changes in toxic releases within facilities between 1988 (the first year of TRI reporting) and 2007 (U.S. EPA 2009). Overall trends are impressive, with a drop from 3 billion pounds of releases (counting only core chemicals) to 1.2 billion pounds. Specific facilities offer an even more dramatic picture. Between 1988 and 2007, Cytec Industries, located in Jefferson County, Louisiana, decreased its releases by 70 percent, dropping from 28.4 million pounds to 8.4 million pounds. Similarly, Eastman Kodak in Monroe County, New York, went from 18.1 million pounds of toxics to less than 1 million pounds, a drop of almost 95 percent in total releases. Some of the most significant improvements occurred in the earliest years of the program, between 1988 and 1991. For example, Invista, a facility in New Hanover County, North Carolina, reduced its toxic releases from 25.6 million pounds in 1988 to 3.6 million in 1989—a drop of roughly 85 percent in only one year's time.

In part because of these early successes, in February 1991 the EPA began a related initiative, the voluntary 33/50 Program that sought to decrease industrial releases of 17 high-priority (highly toxic or widely used) chemicals by 33 percent below a 1988 baseline by the end of 1992 and by 50 percent by 1995. The EPA eventually invited some 8,000 companies to participate in the program, and about 1,300 did so, representing more than 6,000 facilities. The EPA reported that the companies succeeded in reducing toxic releases of the 17 chemicals by 46 percent by 1993 and 60 percent by 1996. Critics have long observed that some of the reported reductions actually took place before the program began, some were made by companies that were not participants in the program, and some were probably not directly attributable to the program itself. Still, the relative success of the initiative boosted confidence in what voluntary pollution prevention programs might be able to achieve (Coglianese 2010; Dahl 1997; Gamper-Rabindran 2006; O'Toole et al. 1997; Press and Mazmanian 2010).

Based on seemingly favorable improvements in environmental performance, the TRI program in particular has been much praised, by the U.S. EPA, state and federal policymakers, students of public policy, and environmental groups (such as the Environmental Defense Fund and OMB Watch) that have made the database widely available. The data also have been used to identify and respond to environmental justice issues since they facilitate an understanding of how toxic chemical releases and risk levels vary from one community to another, including communities with large concentrations of minorities or low-income populations. Policy scholars continue to cite the TRI program as perhaps the premier example of a federal nonregulatory environmental policy that has worked fairly well (Graham and Miller 2001; Hamilton 2005; Press and Mazmanian 2010).

In this chapter we review the record of the TRI's achievements as well as its shortcomings. We report on the quantitative analysis of our data, which focuses on changes over time in our sample of facilities that release TRI chemicals, using several different ways to measure environmental performance. We explain our key dependent variables and why they were chosen. They include indicators of changes in both releases of toxic chemicals and their risk levels as measured by the EPA's Risk Screening Environmental Indicators (RSEI) model.

As a preview of the chapter's argument, we find that on average facilities have indeed lowered the amount of their toxic releases, but there is a quite varied pattern across the nation—both from state to state and from one industrial facility to another. Some states and facilities have made much more progress toward pollution reduction goals than have others. We believe this finding is insufficiently emphasized in most discussions of the TRI program, and by extension in review of other information disclosures policies and other kinds of voluntary corporate action, such as adoption of environmental management systems and related forms of certification of social responsibility (Prakash and Potoski 2006). It is for this reason that we want to examine both green and brown facilities and ask about the differences between them and how we account for this striking variation.

Another key finding is that somewhat contrary to the expectations that policymakers had for the TRI program, relatively few people and community groups have made much direct use of the TRI data; this is particularly the case in later years of the program's operation. Similarly, media attention to the annual TRI releases appears to have dropped off sharply since the early reports of the late 1980s and early 1990s.[1] In the early days it was not unusual for TRI reports to be covered prominently, particularly in local newspapers, as the information was seen as new and somewhat controversial. This seems rarely to be the case today. Our impression of this change in media coverage is reinforced by our national survey of manufacturing facilities, the results of which we describe in chapter 5. Few of our respondents reported that they heard much from the local communities or from the local media. Indeed, the evidence indicates that community pressure does not seem to be a driving force behind chemical management decisions. Rather regulation and concern about potential financial liability more strongly affect corporate decisions about chemical management. Despite this pattern, we will argue that the TRI has had a greater impact than one might suppose, albeit an indirect one that is related more to industry knowledge of its chemical releases and responses to changing public expectations for corporate environmental behavior as well as the use of TRI data by state and federal regulators to improve their oversight of chemical management.

Whatever the reasons for corporate actions, the overall pattern of reduced chemical releases by TRI facilities, so often emphasized in appraisals of the program, is one side of the toxic chemical coin. The other is that the nation's facilities continue to release a prodigious quantity of chemicals with somewhat uncertain but nonetheless worrisome effects on public health. As we reported in chapter 1, the 2007 TRI Public Data Release (U.S. EPA 2009) puts the total national releases at 4.1 billion pounds of toxic chemicals that are disposed of or released on-site or off-site (mostly, 87 percent, on-site).[2] Persistent bioaccumulative toxic chemicals (PBT) accounted for 12 percent of the total, reflecting some 507 million pounds. Moreover, the report also tells us that some 835 million pounds, about 20 percent of the total releases, consisted of 179 known or suspected human carcinogens.

These numbers suggest that the nation could and should do better in managing its toxic chemicals and their release to the environment. Such a conclusion is reinforced by reports during 2009 of persistent, widespread, and serious problems related to toxic chemicals in the nation's surface water as well as in municipal drinking water.[3] We think our analysis suggests some ways to improve the way in which information is collected and disseminated to the public, a subject to which we return in chapter 7. Such information provision is surely one of the more important components of governmental efforts to improve environmental quality nationwide and thereby to reduce public health risks related to exposure to toxic chemicals.

Evolution of the Toxics Release Inventory

In chapter 1 we briefly recounted the history of the TRI program to introduce it and place it among other information disclosure programs within the United States. We elaborate on that history here to show not only how the TRI came to be, but also why it took the form that it did and what difference this policy development has made for TRI's operation, achievements, and shortcomings. As noted earlier, the TRI program's origins may be traced in large part to the tragic Bhopal chemical plant disaster of 1984 and public responses to it. There is no question that the global reaction to Bhopal was immediate and strong in the United States as it was elsewhere around the world. People were angry to learn that corporate facilities could pose such an enormous risk to public health without knowledge of that risk within the local community, and they feared that other comparable accidents could occur, also without local knowledge of the chemical risks. Policymakers and other political actors both in the United States and around the world saw the lack of such corporate transparency as a problem that needed fixing.

The timing of the Bhopal accident was fortuitous for such fixing within the United States. Key members of Congress at the time were considering changes in the federal Superfund program, and new concern over the public's right to know about toxic chemical risk could be linked easily to that program's renewal. The result was the adoption of SARA and its new Title III, the Emergency Planning and Community Right to Know Act (EPCRA) that created the TRI program, as discussed in chapter 2.

In a proximate sense, the Bhopal disaster was a motivating or catalytic event that made the TRI program possible. Yet to start with the Bhopal incident misses the wider context and history of a program such as the TRI. As a form of information disclosure policy, its origins are much older. In particular, it can be traced back to concerns about financial reporting following the stock market collapse of 1929 and the economic depression that followed. At heart, the issue concerned the obligation of publicly-traded corporations to make financial information publicly available. In 1934, the U.S. Congress enacted the Securities Exchange Act, which requires that public companies report regularly to the U.S. government about their financial well-being. In turn, the government makes this information available to the wider public. This policy action reflected a belief that lack of transparency about the financial conditions of publicly owned firms contributed to the market crash and the financial ruin that followed. Policymakers believed that public disclosure would reduce risk, both to the individuals investing in these firms and to the wider economic system. The firms were wary of having to reveal too much about their business practices, but the prevailing sentiment was that government had to require the release of such information and had to monitor the process to promote the public welfare.

The connection between financial reporting requirements dating to the 1930s and similar mandates today that affect the use and release of toxic chemicals concerns risk and risk management. In order to avoid further disasters, government would force companies (or in the TRI's case, facilities) to report information that would otherwise remain private. This information could then be used by others to understand a variety of risks that might affect them, such as industrial accidents or the routine release of chemicals to the environment. Much in the same way that investors had a right to know some basic information about the firms in which they were investing, community members had a right to know about chemicals being stored or the pollution being produced at nearby industrial facilities.

Even when such rights are widely recognized, policymakers do not always develop new policies. This is particularly the case when industry is likely to resist the imposition of additional regulatory requirements or other costs and burdens and when the saliency of the issues remains low for the general public. Given the influence of business in environmental policy, special circumstances must exist to overcome such resistance and to adopt new programs (Kraft and Kamieniecki 2007). So how do we explain the development of EPCRA in 1986?

One useful lens that can help to explain this development and put it into a larger political context is John Kingdon's (1995) model of agenda setting and policy change. Kingdon argues that three largely independent streams of activities can be thought to flow through the political system at any point. The three streams of his model—the problem stream, the policy stream, and the politics stream—flow and shift at their own pace in response to various social, economic, and political changes. At some point, however, the three streams converge when a window of opportunity arises and policy entrepreneurs are able to promote particular policy strategies because of the level of attention and political support found at such times. All three streams were clearly relevant when the Congress finally enacted EPCRA, and the Kingdon model of agenda setting and policy change seems to fit well in this case. It helps to explain how the TRI program emerged and took the form that it did.[4]

The overarching *problem* was the potential for negative health effects attributable to exposure to toxic chemicals. Although there were, and continue to be, a host of ambiguities surrounding exposure to toxics and their effects, the central problem has been clear for some time. Exposure to toxic chemicals has the potential to seriously impair human health. Toxic chemicals can be found throughout modern society, including in most homes, but industrial facilities are an obvious focal point. Furthermore, the likelihood of human error and accidents in the handling of toxic chemicals meant that there was a recognizable risk for exposure by at least some members of U.S. society. The Bhopal accident, followed by the scare within the United States at another Union Carbide facility in West Virginia a few months later, only helped to further crystallize the issue.

The *policy stream* (that is, available policy solutions) for toxic pollution exposure, included the idea of right-to-know requirements, and it had been circulating among policy elites at the federal and state and local levels for a long time. As argued just above, this right to know has been used in a variety of other policy areas, most notably financial disclosure requirements for public corporations, but also campaign finance regulation, occupational safety and health regulation, and many other aspects of social and political life, as we described in chapter 1. As we noted in chapter 1, even before the idea of applying the right-to-know concept came to be recognized as legitimate at the federal level, states and localities had long experimented with it and many had taken legislative action under the guise of "community right-to-know" regulations (Hadden 1989; Lynch 1989).

The support for community right-to-know legislation can be traced back to an equivalent push for workers right-to-know legislation that would help inform workers about the nature of the hazardous substances they were handling or being exposed to in the workplace. Much of the activity around workers right-to-know policy during the late 1970s and early 1980s also occurred at the state and local levels. States such as California and New Jersey as well as cities such as Philadelphia enacted legislation to assist workers in this manner. In the early 1980s, attempts were made to broaden right-to-know provisions to include communities that surrounded industrial facilities (Hadden 1989).

By 1983, the Occupational Safety and Health Administration (OSHA) had taken up the issue of workers' right to know about chemicals in the workplace and later in that year it pushed forward a rule that required all industrial facilities in all states to provide information about toxic hazards on the shop floor. In some respects, this was a benefit to industry, which would face uniformly applied federal regulations rather than disparate requirements across the fifty states. Nonetheless, OSHA's rule helped to set the stage for what was to come.

Given the reaction in the media and among citizens worldwide to the accident in Bhopal, the role of the *politics* stream in helping to create the TRI program is relatively straightforward. As we discussed in chapter 1, in 1985 the U.S. Congress took up the question of how to best revise and update the Superfund program for cleanup of hazardous waste sites. Concerns about liability requirements and the program's somewhat shaky early implementation (Cohen 1984) led some legislators to seek revisions of the law, and policy entrepreneurs in Congress, such as Representative James Florio, D-N.J., and others outside of Congress, used the Bhopal tragedy as a springboard for federal legislation on information disclosure. Even the chairman of Union Carbide, the company that owned the Bhopal plant at the time, said in an interview a month after the explosion, that right-to-know legislation at the federal level in the U.S. was inevitable. The chemical industry began developing a multifaceted response to new public concerns and the likelihood of federal policy action (Chemical Week 1985).[5] In Kingdon's terms, the politics stream made action almost certain. The idea was increasingly popular, policymakers thought protection of the right to know about toxic chemicals was a legitimate federal responsibility, and industry was unlikely to succeed in blocking such measures. The reauthorization of the Superfund program became a perfect vehicle from the policy stream for moving forward with right-to-know legislation.[6]

Although the final legislation took a good part of a year to assemble and revise to meet varying concerns, including differences between the Senate and House versions, by 1986 a compromise was struck and SARA was enacted into law and signed by President Ronald Reagan. Although EPCRA and the TRI program that SARA created have been faulted in some critical appraisals over the years, it is fair to say there has been no serious effort to repeal the legislation; indeed, as we discussed in chapter 1, changes made by the Bush administration in 2007 to reduce TRI reporting requirements were quickly reversed by Congress and the Obama administration in 2009, showing strong political support for the program. This support continues despite arguments advanced by some critics that the collection and release of TRI data are not as useful as once the case (Atlas 2007). These conditions reinforce our interest in learning more about just how effective the TRI program has been and the mechanisms through which it achieves its objectives, to the extent that it does.

TRI Program Requirements and Operation

The basic operation of the TRI program is quite simple. Sections 311 and 312 of EPCRA require that industrial facilities covered by the law report annually to state and local governments using forms created by the EPA on the locations and the types and amounts of toxic chemicals released to the air, water, and land and to account for the quantities of toxic chemicals sent to other facilities for waste management. Facilities have to specify the location from which the release has occurred, e.g., on-site or off-site, and provide the information chemical by chemical. Section 313 of the law mandates that the EPA and the states collect data on the releases and transfer of the listed toxic chemicals and make the information available to the public via the inventory, which has been available on-line since the late 1990s. In addition, as noted previously, the 1990 Pollution Prevention Act requires that additional data on waste management and source reduction be reported.

The EPA continues to remind readers in its Web site description of the program that its goal is "to empower citizens, through information, to hold companies and local governments accountable in terms of how toxic chemicals are managed." Armed with the TRI data, the agency says, communities "have more power" to make informed decisions, and companies are spurred "to focus on their chemical management practices since they are being measured and made public." In addition, the EPA says, the data can serve as a "rough indicator of environmental progress over time."[7]

Contrary to what many people may believe about the TRI, the law does not require measurement of actual chemical releases; rather, facilities may report estimates of their releases. Most facilities have no reliable way to measure releases and most of the data reported to the EPA and the states are estimates rather than measurements. Those who complete the TRI forms at the facilities (sometimes employees and sometimes staff at consulting companies who specialize in this process) depend to some extent on the previous year's reports, making adjustments as needed. As is the case with other environmental policies, facilities that misrepresent their releases are subject to penalties.

In its early days, the TRI identified some 300 chemicals that needed to be reported. Over the years, that list has changed in many ways. Most notably, the number of reportable chemicals has grown substantially, and by 2009 totaled nearly 650 chemicals. Some chemicals on earlier lists have been removed, but many more have been added. The kinds of industrial facilities that are required to report under the program also have expanded over time. Seven new industrial sectors were added to the original covered industries since the program's inception in 1987, including metal and coal mining and electric utilities (U.S. EPA 2009).

The EPA has established various threshold requirements for reporting, in part to distinguish larger and smaller facilities and thus greater or lesser risks to a community. Under EPCRA Section 313, facilities must report releases if they have 10 or more full-time employees or the equivalent; they are in a covered industrial code; and they exceed any one of the thresholds for manufacturing, processing, or otherwise using a listed chemical. The list is provided in the Code of Federal Regulations (40

CFR Section 372.65), and is updated periodically. The agency also allows for a more abbreviated report via a Form A (instead of the normal Form R) to reduce the burden on facilities. That option is available if the facility meets the employee, industrial code, and chemical activity thresholds that the agency allows. Congress specified in 2009 that the more detailed Form R must be used to report all persistent, bioaccumulative toxic (PBT) chemicals, but the EPA still permits the Form A to be used if the facility is reporting 500 pounds or less each year and less than one million pounds of the chemical were manufactured, processed, or otherwise used during the reporting year.[8]

As this description makes clear, the TRI program does not capture information about all toxic chemical releases from all manufacturing facilities. Rather it is intended to provide information about most industrial facilities that release substantial quantities of the listed chemicals, presumably because these releases expose communities to chemicals that pose the greatest public health risks.[9]

Beyond the process of consolidating the information that is reported and making it available to the public, and revising the program's requirements as needed, the EPA also offers assistance to industry. This assistance comes through provision of a variety of training sessions that can be particularly helpful to smaller facilities without the experience or expertise that are common with larger manufacturing enterprises. In one of the more consequential recent actions, particularly from the industry perspective, the EPA developed online reporting capabilities that can help to reduce the time (and thus the cost) associated with TRI reporting, a long-standing complaint by facility managers.

Industry Responses to the TRI Program

Over the life of the program, industry responses to the TRI program have been mixed (Santos, Covello, and McCallum. 1996). Before Congress enacted EPCRA, some firms and business groups fought the idea, as they have many other proposals for new or expanded federal and state environmental policies (Eisner 2007; Fiorino 2006; Kraft and Kamieniecki 2007). They argued that the program would put undue financial strain on companies by forcing them to collect, organize, and report toxic release at their facilities. Others wondered if the program was an efficient use of limited resources, given the many other pollution control issues that firms needed to address (U.S. Senate 1985). The most common response, however, was lack of attention to the legislation. Other bills being considered at the time, including updates to the Superfund program, garnered much more interest overall.

Despite the limited attention given to the program in its early stages, industry was able to influence its development, if only at the margins. For example, the EPA modified its requirement of facility data accuracy. Rather than requiring a signed certification of accuracy from senior management, the language was softened so that management was not required to certify that it has conducted a "personal examination of the completed forms" (Hamilton 2005, 49).

Yet industry also had to accept decisions that it would have preferred not be included in the legislation. For example, the EPA decided not to include risk information with the data gathered from facilities. Instead, the agency left it up to the users of the TRI information to decide how to interpret the risk from different amounts of chemicals. The decision is somewhat surprising because of the obvious difficulty of making such determinations, especially on the part of ordinary citizens (Hamilton 2005, 51). The decision to include thresholds for reporting (as noted above) also was made at this time.

In the first years of the program, much of the attention by industry was given to making sense of their own reporting results at the facility level and comparing their toxics release levels to those of similar facilities and industries. The EPA heard complaints about the lack of clarity in agency reporting requirements and the frustration facilities faced in trying to figure out how best to estimate their release levels at the level of detail required by the agency. At the same time, some companies were alarmed by the results they were finding at their own facilities and voluntarily committed themselves to reduce their release levels in subsequent years. The chemical manufacturing industry, in particular, took significant steps toward release reductions, recognizing the value of improving its public image through such an action. In part for this reason, as we discussed in chapter 1, the industry was moved to adopt its Responsible Care program as a highly visible effort to demonstrate that its member firms cared about the effects of manufacturing operations on the communities in which they were located (Press and Mazmanian 2010).

Some facilities paid special attention to loopholes in the reporting requirements, thereby making "paper changes" to their reported pollution output (Natan and Miller 1998). Although the vast majority of facilities attempted to be accurate in reporting releases, a smaller percentage appeared to game the system as best they could.

As the program developed over time, facilities raised two related concerns. One was the challenge of understanding and responding to new reporting requirements, which could, and usually did, change from year to year. The second was that the requirements were excessive or too detailed, and they sought changes that would be more supportive of business. Many companies came to see data collection for the TRI as a rote process, requiring a bit of stamina and the ability to deal with tedious details. As is the case with so many other activities in the corporate world, a good deal of the data collection, compilation, and reporting has been outsourced to a cottage industry of TRI reporting experts; some companies handle the tasks in-house while others farm out almost all of them to consultants. Even if this kind of adjustment to yet another routine reporting requirement for business has become commonplace, some facilities and companies continue to complain that the EPA's requirements are burdensome. The burden grew in part, they said, because the agency was constantly making adjustments to the program, which meant that in turn the facilities had to learn something new about the reporting requirements, adding to the time and cost of TRI activities.

Two substantial changes occurred in the mid-1990s that were not at all well received by industry. First, in 1995 facilities were required to almost double the number of chemicals for which they were to report; the number jumped from around 300 to close to 600 in just one year. As is often the case in responding to new regulatory requirements, facilities argued that the extra paperwork and the data that would be reported were not worth the considerable additional cost to industry.

The second change came in 1998, when mining companies, electrical utilities, and a few other previously exempt industries were required to report their pollution releases. Some of these industries fought their inclusion, partly because they understood that when required to report, they would rise to the top ranks of polluting industries within the TRI program and attract unwelcome attention by the media. This program change also altered the ranking of states. For example, a state like Texas, which spent much of the early and mid-1990s at the top of TRI-polluting states because of its large industrial base, dropped further down on the list of the most polluting states; it was replaced by those states with extensive mining operations, such as Nevada and Alaska.

By the late 1990s, the process of reporting TRI data became more straightforward for many facilities, yet the EPA continued to receive regular queries about the reporting requirements. With this in mind, EPA staff developed TRI-ME, interactive software that allows facilities to enter data into the program more easily than earlier, increases the accuracy of reporting, and reduces required paperwork. Originally rolled out in 2000, TRI-ME was subsequently improved to better meet the needs of those organizing and sending data to the EPA. As we will report in later chapters, our survey of industry TRI contacts and corporate interviews reinforce the success of these efforts. Many individuals told us that the process of reporting TRI data in recent years had become much easier and thus was less of a concern to them than it had been in previous years.

Parallel Regulatory Changes in the States

Changes in the TRI program over the span of the 1990s did not take place in isolation. Rather, at state levels, new policies were being developed and programs were being shaped with an eye on improving environmental performance. New Jersey was leading the way when it came to materials accounting. Starting in 1994 and running to the present, the state's Worker and Community Right to Know Act requires private industry to provide comprehensive documentation to the state about chemical use and release. Toxics use reduction was enshrined into law in Massachusetts in the mid-1990s as well. The state's Toxic Use Reduction Act had requirements similar to New Jersey's materials accounting, plus other components such as the creation of an advisory board to oversee the new law and requirements that facilities in Massachusetts submit environmental management plans to the state for its review.

Programs in states such as New Jersey and Massachusetts suggest that federalism has been alive and well in the area of toxic chemical management. State policy and its determinants have received more attention because of growing state responsibilities (Nice and Frederickson 1995), the devolution of domestic policy power (Van Horn 1996), and the expansion of state government capacity (Weber and Brace 1999). By the end of the 1990s, an era of "state-centered policymaking" (Hanson 1998) or "state-based government" (Donahue 1999) was widely recognized. States became ascendant in this reconfiguration of policy responsibilities between them and the federal government (Nice 1998; Rabe 2010).

As chapter 4 will show, state-level variations in environmental performance are partly explained by state-level political, economic, and administrative factors. A comprehensive analysis of national trends appropriately includes a deeper analysis of variations at multiple subnational levels, and we provide that in chapter 4.

Measuring Performance

The understanding of TRI's purpose has changed over time. In the beginning it was meant to serve as a supplement to command-and-control regulation in order to deter bad behavior on the part of industry (Hamilton 2005). Citizens had a right to know about potential harms due to toxic releases and the TRI would partly serve to make them aware of the risks around them. To use the language of chapter 2, the program designers were in the more traditional position of being constrained by the "regulation dilemma." The TRI mandated disclosure, but partly with the idea that transparency may lead to voluntary attempts to reduce releases (Potoski and Prakash 2004). Over time the program has been understood to also draw attention to environmental performance.

Much like other environmental policies, TRI's ultimate goal can now be understood to be to improve industry's environmental performance as measured by reduced release of toxic chemicals; such a change in turn should lower the risks to public health presented by those chemicals. The desire for cleaner water, land, and air drove the TRI's creation, and reductions in releases and risks speak to this desire. As we stated in chapter 2, assessments of any of these policies, including the TRI program, should focus on the extent to which they have achieved such goals, and in chapters 4, 5, and 6 we will turn to the mechanisms through which information disclosure under the TRI program has made a difference. Here we concentrate on the key dependent variables for environmental performance: changes in releases and risk levels over time. We explain their selection and measurement, and we survey the macro trends in industry's performance since the late 1980s.

A distinguishing feature of the TRI program is that facilities are to report on the number of pounds of chemicals being released to the environment. Asking for data about *pounds* of pollution might seem a little esoteric and perhaps too detailed for many people. People are concerned about exposure to toxic chemicals but they also may have trouble understanding how a certain number of pounds of releases translates into their personal exposure and thus to what extent they ought to be concerned about health effects. Moreover, even though by definition TRI chemicals are "toxic," some are far more toxic than others. A pound of one chemical or chemical compound may be quite toxic. In contrast, a pound of another chemical may not be nearly so dangerous to public health even if thousands of pounds are released from a facility.

A further complication in seeking to learn about TRI release trends over time is that, as noted earlier, the EPA has expanded its list of chemicals over the years. During six years of the program's first 17 years, the list changed. New chemicals were added, while others were dropped off the list. In 1991, 1995, 1998, 2000, and 2001, new chemicals were added to the list. Throughout this book our standard practice will be to use what is called 1991 core chemicals in describing changes in pollution releases over time. Obviously following that convention means that we are not tracking the newer chemicals added to the list since 1991 and thus the complete environmental performance of facilities that release these chemicals. At the same time, to have a meaningful metric for facility performance over time suggests the wisdom of sticking with the original or core list of chemicals, and this is the way that the EPA and many others report progress on chemical releases from the late 1980s to the present. When reporting data for other years, we will specify any change in that conventional usage.

The TRI program requires facilities to split out their data by the medium through which releases occur. Releases to air, water, and land are likely to have different impacts and require different assessment. Air pollution dissipates more quickly than other forms of pollution, yet it is also the medium by which many people are exposed to toxics. Inhalation is a primary means by which toxic chemicals can cause harm. Water pollution can also be distributed widely, but it may be concentrated in certain places. Ingestion of contaminated water is one conduit for environmental harms. Depending on the nature of the chemical, water contamination can remain high for some time. Exposure to toxics in water can occur through dermal contact and through ingestion of food that has been exposed to toxics (for example, fish that absorb toxics, such as mercury or polychlorinated biphenyls (PCBs), into their bodies). Land pollution is both the least likely pathway for exposure to toxics and yet is also often the form of contamination that lasts the longest. Dermal contact is the most likely form of exposure, with the possibility of ingestion as well. Soil contamination can last decades, if not centuries, but it can also be contained much more easily than either air or water pollution. Of course, some land disposal practices may also have implications for contamination of groundwater or surface water bodies just as some air releases will find their way into water bodies. For example, many lakes and ponds are contaminated with mercury, the sources of which are hundreds or even thousands of miles distant from those water bodies; the contamination comes from air deposition of mercury.

Further distinctions are made in TRI data. The difference between on-site and off-site disposal can be substantial, if only because where the toxics can be found has a direct relationship to which populations have a chance of being exposed to those toxics. Toxics can be disposed of in landfills, through surface water discharges, and through underground injection, among other actions. Each of these possibilities is likely to have an influence on the risk created by the releases because different waste management choices have different impacts on the surrounding environment.

Some Caveats about the Data

It is important to keep in mind a few other distinctive features of TRI data. First, the TRI covers only a subset of all possible sources of pollution. In particular, industrial point sources make up the bulk of the organizations covered by the program. Historically that has included major industrial sources such as chemical companies, pulp and paper operations, plastics manufacturers, and metal companies. In more recent years, as noted, the TRI has been expanded to cover metal and coal mining and electric utilities.

Second, the TRI covers only a subset of all possible kinds of pollution. Numerous chemicals are not on EPA's list for required reporting and in many cases we do not know much about their toxicity. The current list, for example, while much larger than it was in 1987, still contains only about 650 chemicals. Yet there are tens of thousands of chemicals used in commercial quantities in the United States and perhaps a thousand new ones are produced by industry each year. Most of those chemicals are considered to be safe, and yet they have not been thoroughly tested for toxicity. Indeed, in the 1980s there were essentially no data on the toxicity of over three-quarters of the chemicals that were most commonly used in commerce, that is with 1 million pounds or more of each produced annually (Shapiro 1990; Sigman 2000).[10]

Third, when we turn our attention to our subset of facilities that reported in 1991, 1995, and 2000, it is important to recognize that this means we are not including facilities that reported in only one or two of those three years or that did not report in any of these years, despite having reported in other years. We are therefore looking more closely at a relatively stable subset of facilities. They are stable both in terms of their continued existence during this time period and because they have maintained some basic level of industrial production which leads to their need to report toxic pollution data.

Finally, and probably most importantly, TRI data are self-reported. Although the EPA can evaluate facilities for their compliance with the law, in the bulk of cases no such screening is performed. Some companies do not report when they should, mostly it seems, due to misunderstandings about the program's requirements (de Marchi and Hamilton 2006). The fact that the data are self-reported also raises some questions of accuracy. At least two investigations by third parties in recent years have suggested that the EPA's data are not as accurate as they might be (Environmental Integrity Project 2004; de Marchi and Hamilton 2006). These investigations follow previous analyses that indicated at least some of the early drops in releases levels were due to "paper changes" rather than true changes in production or chemical use (U.S. GAO 1991; Natan and Miller 1998).

Why Focus on Facilities?

It is important to note that the flexibility of the TRI allows for analyses beyond the facility level. TRI data can be aggregated based on the kinds of chemicals being released, trends over time, geographic variables, and industry types. Arguably all levels of analysis are important for increasing our understanding of TRI data. Yet the decision to focus on the facility level is driven by a belief that it is here where changes in environmental practices occur, and therefore here where changes in releases over time deserve the most attention.

There is at least one other level of analysis not covered by TRI data; this is firm-level data. Environmental management practices at the firm level are connected to facility level performance in many if not most cases. It would be a mistake not to acknowledge this reality. Yet because the EPA requires facility-level data, and neither collects nor distributes firm-level data, firms are not as likely to receive scrutiny across the entire corporate entity, but rather based on what happens at particular facilities. Facilities are the locus of attention.

Facilities are also quite tangibly the places where changes can occur. If a firm wants to reduce its toxic chemical releases, it must deal with each facility it owns to figure out the best course of action. Changes in equipment or chemical use are driven by the particular context of a given facility, including which state the facility is located in, when the facility was built or last updated, what is being produced there, and the capacity and motivation of facility-level management.

For our purposes, the focus on facility-specific chemical management also makes sense because it is at this level that we would expect some consideration to be given to the surrounding community, and in turn one would expect community residents to take a special interest in the management and releases of the local facilities even if key decisions are made at a corporate headquarters. In our survey of TRI facility officials, we specifically asked about the relationship between corporate headquarters and facility management, and we address that subject in chapters 5 and 6.

Why Concentrate on Air Releases?

Although we do not focus exclusively on air releases, it is these kinds of release of chemicals to the environment to which we give the most analytic attention. There are several reasons for this. One is that air releases make up the bulk of all chemical releases reported under the TRI program. Between 1991 and 2000, air releases made up an average of 58 percent of all releases per year among 1991 core chemicals. Over this period of time, air releases ranged from 69 percent (a clear majority) of all releases to 47 percent (a plurality). The general trend was downward over this period of time, suggesting that reductions in air releases have moved more quickly than reductions in water or land releases.[11]

Another reason is that air releases tend to create the most controversy. As mentioned earlier, air pollution tends to affect the most people and it is dispersed quite rapidly. Exposure to air releases—usually through inhalation—can more directly and severely influence human health than is the case with releases to water or land. Air releases tend to be more obvious and are therefore perceived (rightly or wrongly) as more consequential.

Finally, as described below, we are able to go beyond air releases in pounds of chemicals to look at human health risk through use of the RSEI model. Equivalent data are not available for land releases, and although recently data have become available for water releases, they are not as reliably modeled for toxic exposure.

Going Beyond Releases: Measuring Public Health Risks

In the years since the EPA began the TRI program, the agency has increased the environmental information it makes to the public, thereby increasing the ways in which environmental performance can be evaluated. There are at least three new forms of information which have gained some attention at different times during the life of the program.

First, data are now available on what some researchers refer to as "production related waste" (PRW). As we noted in chapter 1, PRW is the sum of all toxic wastes generated across a firm's production processes that a facility reports as recycled, recovered for energy, treated on and off-site, or released on and off-site. TRI releases can be understood as a subset of PRW. Though TRI data are a more precise measure of releases, PRW data are a better measure of all toxic wastes produced at a facility.

Second, over time the EPA has gradually expanded its disclosure of enforcement data. Facilities can be penalized for not meeting federal regulatory requirements. The nature of these penalties, including the kind of violation and the amount of money the facility has been fined, are now publicly available in many cases.[12]

Third, in recent years, new models for risk have been introduced. Much has been made about what the TRI does not disclose and, in particular, the lack of any risk characterizations that would allow a comparison of various toxic releases (Bouwes, Hassur, and Shapiro 2001; Graham and Miller 2001). In fact, EPA documentation on using the TRI begins by telling potential users that the database's chemicals can vary widely in their toxic effects. One's perception of and attention to high-volume releases may be misdirected when more toxic chemicals are being released at lower volumes (U.S. EPA 2002, 2009). As one group of researchers noted, "the human health impacts of the various carcinogens and noncarcinogens in the inventory can differ by up to *seven and eight orders of magnitude*, respectively. That is, a single pound of the most toxic chemicals . . . is toxicologically equivalent to one hundred million pounds of the least toxic of these substances" (Bouwes, Hassur, and Shapiro 2001, 3–4, italics added).

One of the ways of overcoming this limitation in TRI data is by applying a risk model. The EPA's Risk Screening Environmental Indicators (RSEI) model performs such a function. The RSEI software begins with the chemical and its air release amount and puts it into a steady-state Gaussian plume model. It then simulates downwind air pollutant concentrations from a stack or fugitive source as a function of facility-specific parameters (e.g., stack height and exit gas velocity), local meteorology, and chemical-specific dispersion and decay rates. These factors are then overlaid on demographic data taken from the U.S. Census to produce a surrogate dose estimate for the surrounding population. The final product of applying the RSEI model is an indicator value that represents a risk characterization that permits users to discern and compare chemicals with dramatically different toxicological effects that are released from manufacturing facilities.

The RSEI model allows us to estimate public health risks associated with a given facility over time in addition to its reported releases of chemicals, giving us two interrelated measures of environmental performance. In this way we can construct a new dependent variable that reflects the relative performance of a given facility along these two dimensions: releases and risk. Facilities can be placed within a two-by-two matrix that distinguishes high and low performance levels along each dimension. *Green* facilities are those that saw decreases in both releases and risk levels. *Blue* facilities are those that saw a decrease in risk levels, but an increase in release levels. *Yellow* facilities are those that saw a decrease in release levels, but an increase in risk levels. Finally, *brown* facilities are those firms that saw an increase in both releases and risk levels.[13] These four types of facilities are shown in table 3.1.

We chose to study facilities using these categories rather than relying on raw data alone. One reason for doing reflects our discussion earlier of the limitations of TRI data. Although one can challenge the accuracy of both release and risk data, there is much less contention about whether

Table 3.1
Typology of Facilities

Risk	Releases Increasing (Dirtier)	Decreasing (Cleaner)
Decreasing (Safer)	Blue facilities Example: a firm could substitute a more benign chemical for one of its most toxic air releases, but still generate and even release large quantities of less toxic pollutants.	Green facilities Example: a firm installs new pollution control equipment that decreases the volume of its more toxic air releases and initiates source reduction activity that reduces its risk levels.
Increasing (Riskier)	Brown facilities Example: a firm increases production but takes no steps to control the higher volume of toxic air releases and the risk they pose.	Yellow facilities Examples: a firm targets its biggest releases for reductions while maintaining or even increasing a low volume, but highly toxic (riskier) air release.

facilities have increased or decreased their release and risk levels. Our categories allow for a conservative, more reliable measure of change. Another reason is that our measure captures the extent to which all facilities are seeing positive or negative change in their release or risk levels. The raw data skew the analyses toward those facilities producing the greatest quantities of hazardous releases. Indeed, it has long been the case that a relatively small number of facilities are responsible for a disproportionate share of the chemical releases, so it is important to take that pattern into account. We discuss the data later in the chapter. Our categories allow us to better evaluate whether TRI-reporting facilities in general, and not just the largest ones, are moving in a more sustainable direction.[14]

Environmental Performance at the Facility Level

Air Releases
Our database includes the performance of the 8,389 facilities that reported via the TRI in 1991, 1995, and 2000. Among these facilities, the amount of toxic air releases was reduced significantly. The facilities went from a combined level of 1.21 billion pounds of air pollution in 1991 to 965 million pounds in 1995, and to 677 million pounds by 2000, an eventual decrease of 44 percent (see figure 3.1). This average decrease is based on facilities that reduced their air pollution during this

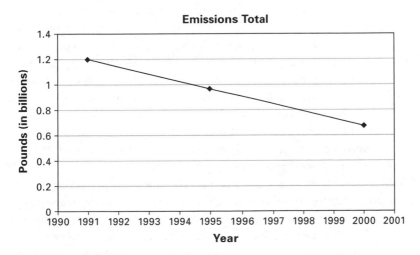

Figure 3.1
Change over Time in Toxic Emissions Totals

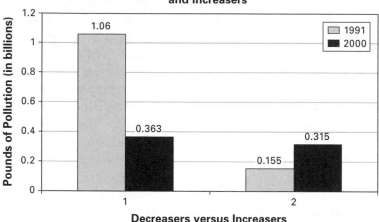

Figure 3.2
Changes in Emissions Totals for Those Facilities Decreasing or Increasing Their
Emissions Levels between 1991 and 2000

period, those that increased it, and those reporting no change. These are
the kinds of figures that have attracted so much attention and seemingly
point to the great success of the TRI program in reducing the release of
toxic chemicals across the nation. But the overall trend is only part of
the story, and as is often true of statistical averages, the numbers disguise
substantial variations among facilities and among the fifty states.

To focus first on variation among facilities, we can separate out the
two groups of facilities that reported either a decrease or increase in
release of toxic chemicals. Here the trends are even more striking. Of
the 5,213 facilities that decreased their release of toxics, the total amount
of releases dropped from 1.06 billion pounds to 363 million pounds, a
decline of 66 percent. Yet, of the 3,176 facilities that increased their
releases (or had no change), the total amount of releases went from 155
million pounds of toxics to 315 million pounds, a doubling of the
amount of releases (see figure 3.2).

While commentators who address the TRI almost always point to the
average decrease over time in chemical releases, it is equally important
to recognize that the general pattern of improved environmental performance is misleading because it fails to take into account the facilities that
are moving in the other direction. Both observations are important. The
average performance is encouraging and it suggests that the TRI, along
with the effects of other environmental policies as well as independent

changes that are made in facility chemical management, is resulting in greener industrial operations and presumably better health for citizens in the surrounding communities. At the same time, one is struck by the large number of facilities that apparently have been much less affected by these forces and that continue to pose a health risk to some communities. It would be useful to know more about why some facilities become environmental leaders and others are laggards in this regard, a subject to which we return in chapters 5 and 6.

Similarly, when TRI data are reported, our attention tends to be focused on the national trends. Yet facilities are not evenly distributed across the nation, and we would expect to see considerable differences from state to state. For example, the average state had just under 175 facilities within it, but few states were near the average. Rather, some states, such as Alaska or North Dakota, had only a handful of facilities within them, while other states, such as Ohio and Texas, were home to hundreds of facilities. Ohio alone had 8 percent of all facilities within our sample (N = 698) and 5 percent of all releases in 2000. Alaska and Hawaii, at the other end of the spectrum, had only three facilities each and a minimal level of releases. Table 3.2 reports the number of facilities within our sample by state. This variation among states in the number of facilities leads us to believe that states also might vary considerably in the extent to which they affect the performance of industry, and the data strongly suggest this is the case. In chapter 4 we analyze variation across the states in releases and risk levels and try to account for it.

Facilities also are not evenly distributed across industries. Sixty percent of the 8,389 facilities in our sample can be categorized in just four industries: chemicals, plastics, primary metals, and fabricated metals (see figure 3.3). The remaining 40 percent of facilities are scattered across such industries as textiles, lumber, paper making, leather, and printing (and 24 other categories in total). The chemical industry alone makes up just over 23 percent of all facilities under study. On balance the chemical industry was a significant reducer of releases. Of the 1,963 chemical facilities, the average reduction was just over 115,000 pounds of air toxics. The standard deviation was significantly larger than the average, suggesting high variability within the industry. The other three key industries (plastics, primary metals, and fabricated metals) also saw air toxics reductions, although not at the pace of the chemical industry.

Big Players and Bit Players

An analysis of the top 100 in this sample of 8,389 facilities brings up another central characteristic about the nature of toxic chemical releases.

Table 3.2
Number of Facilities within States

Ohio	698	Mississippi	123
Illinois	530	Washington	122
Pennsylvania	524	Oklahoma	103
Texas	509	Oregon	101
California	428	Kansas	101
Indiana	419	West Virginia	70
Wisconsin	377	Maryland	65
Michigan	343	Arizona	58
North Carolina	321	Nebraska	56
Tennessee	265	Colorado	51
New York	260	Rhode Island	48
New Jersey	249	New Hampshire	46
Georgia	241	Utah	43
South Carolina	215	Maine	38
Missouri	206	Delaware	29
Alabama	190	South Dakota	23
Virginia	185	New Mexico	14
Kentucky	183	Montana	14
Massachusetts	179	Nevada	13
Minnesota	167	Vermont	12
Florida	157	Idaho	12
Arkansas	155	Wyoming	9
Louisiana	148	North Dakota	9
Iowa	137	Hawaii	3
Connecticut	137	Alaska	3

Large numbers of facilities in the TRI database are only bit players when it comes to the total level of chemical releases. As noted earlier, a small number of large facilities are substantial contributors to the overall national releases. For example, in 1991, about 33 percent of all reported TRI releases came from just 100 facilities. In comparison, the bottom 7,820 facilities combined contributed only 33 percent of the total releases. About 470 facilities in the middle make up the final 33 percent of the total releases. The top 100 facilities were mostly decreasing their releases between 1991 and 2000 (96 percent of them). The middle group of facilities were also mostly decreasing their releases, but at a slightly lower percentage (85 percent) than the top 100. Finally, the bottom 7,820 facilities were much more mixed, with roughly 60 percent decreasing their releases and 40 percent increasing their releases. Splitting the

Industry Categories

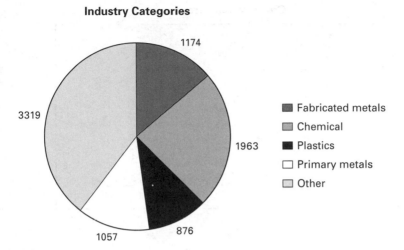

Figure 3.3
Facilities Categorized by Industrial Type

bottom 7,820 facilities in half creates an even more striking picture. Of the bottom 3,910 facilities only 44 percent were decreasing their releases, while of the 3,910 facilities that sit just higher, 75 percent were decreasing releases. These results suggest that where a facility begins may have some relationship to how it can perform over time. We will have more to say about the implications of these results in chapters 6 and 7.

Changes in Risk Levels

Although aggregating risk data for facilities is somewhat challenging because of the nature of the measure, there are several basic points that can be made. First, a significant number of facilities, roughly 55 percent of the 8,389, saw reductions in risk scores between 1991 and 2000. Some of the reductions were minute, but for other facilities they were substantial. Dozens of facilities saw 99 percent reductions in risk. That said, of the approximately 45 percent of facilities that saw risk scores increase, there were quite a few that saw very large increases. It is not unheard of that facilities can increase their risk levels by factors in the hundreds or thousands. Crucial in these latter cases are initial risk levels in the single digits followed by substantial increases in risk levels ten years later.

Second, it is important to reinforce the observation that risk changes and release changes do not always go hand in hand. Although in many cases risk goes up when releases go up or go down when releases go down, contrasting cases do exist. Facilities can see increase in risk at the

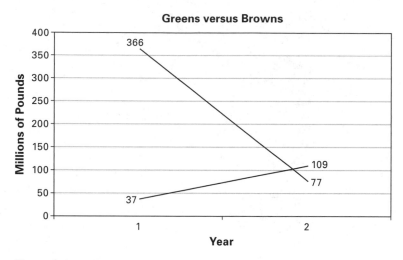

Figure 3.4
Green versus Brown Facilities: Changes in Release Totals between 1991 and 2000

same time that releases go down. That is, facilities can make progress in reducing risk while at the same time they are increasing their overall release of toxic chemicals on the TRI list. Key factors to keep in mind include population changes over time that can affect expected exposure rates and changes in the mix of chemicals that a facility may use, which can greatly influence the toxicity of the aggregate chemical releases.

Third, the number of facilities that are seeing risk levels increase or decrease also varies by industry. Where the general tendency among chemical companies is a decrease in risk levels, the tendency in the plastics industry is, on balance, toward increases in risk levels. Other industries tend one way or the other in a related fashion.

Green and Brown Facilities
When taking a closer look at consistently "green" and "brown" facilities, their release changes highlight the critical influences of these kinds of facilities on overall environmental performance of this nationwide sample (see figure 3.4). Of the 1,447 green facilities, the total amount of release reductions went from 366 million pounds of air pollution in 1991 to 77 million pounds in 2000, a decrease of over 79 percent. Of the 1,082 brown facilities, the total amount of air releases increased from 37 million pounds to 109 million pounds; that is, it roughly tripled in amount.

For risk levels, the numbers are equally dramatic. Green facilities reduced their risk levels, on average, a significant 88 percent between

1991 and 2000. Brown facilities increased their risk levels, some moving from a fractional level of risk to a level that is thousands of times higher.

Looking at the top four sectors for releases (chemicals, plastics, primary metals, and fabricated metals), it should not be surprising that the chemical industry had the greatest number of green facilities at 401 (and more than twice as many green facilities as brown facilities). Both of the metals sectors had roughly the same number of green and brown facilities. There were 130 green facilities in the plastics sector compared to 200 brown facilities.

A Few Illustrations of Variation

Among the 8,389 facilities that we studied closely there are quite a number that made substantial progress between 1991 and 2000 in their pollution reductions. A few, with millions or tens of millions of pounds of toxic air releases in 1991, have decreased their releases by as much as 99 percent, a dramatic improvement that defies reasonable expectations. Many others have decreased their output by one-third or one-half, which is also quite dramatic, given that this translates into millions of pounds of toxic chemicals no longer released to the air each year.[15]

Take, for example, a paper facility in Minnesota. In 1991, the facility produced over 13 million pounds of toxic air pollution. By 2000, the facility was producing only 538 thousand pounds of air toxics, a reduction of significant proportions. The facility's releases in 2000 were 4 percent of its 1991 releases. This is the kind of progress that everyone would like to see, and that U.S. environmental policies, including the TRI, are designed to foster.

Although not as large, other big producers of toxic air pollution also saw significant drops in releases over this ten-year period. A photographic laboratory facility in New York saw a 75 percent drop, shifting downward from 9.3 million pounds of air toxics to 3.2 million pounds. A chemical facility in Texas went from 6.3 million pounds to 3.3 million pounds of air toxics, a drop of almost 50 percent.

The hope behind these large decreases in release of toxic chemicals is that they would translate into cleaner air and lower health risks to citizens in the area. The reality is that in many cases this has quite likely happened, but not in all cases, and in particular for those facilities that are neither consistently green nor consistently brown. Here some facilities are able to reduce their pollution levels, but because of other changes, the health risks to citizens either remain stable or increase. The reasons for this seemingly improbable outcome will be discussed in more detail in the next chapter, but it is important to point out the existence of such cases. The implication for the TRI is that success is not as easily defined as often assumed.

At the same time that there were a number of "success" stories—facilities that improved their environmental performance between 1991 and 2000—there were also those facilities that moved in the opposite direction. A paper company in Louisiana went from just 75,000 pounds of air toxics released in 1991 to over 3 million pounds in 2000. Similarly, an oil company in Texas went from over 660 thousand pounds in 1991 to over 3.6 million pounds in 2000.

Yet not all of the news from facilities increasing their release of toxic chemicals to the air is bad. There is the occasional facility that, despite putting out greater volumes of air pollution, was also moving in the right direction when it comes to reducing risks; that is, risk levels decreased even though the volume of toxic air releases went up. A fabricated metals facility in Arkansas increased its emissions threefold, while diminishing risk by substantial margins. Like the category of facilities described just above, this description speaks to the need for using more nuanced measures of environmental performance. The reality is not always as simple as it seems, and the picture reinforces the argument for looking more at risk levels than total environmental releases. Our concern after all is not just the quantity of material that is released to the air, but what that material potentially does to people and to the environment.

Conclusions

What can be said about toxic pollution releases between 1991, 1995, and 2000? The results described above begin to give us some insight into changes over time in pollution releases. At the broadest level it appears as if the TRI program has tracked a substantial improvement in environmental performance. During a period when the population of the United States was increasing and economic activity grew at a substantial rate, facilities nonetheless have been able to decrease their releases. Yet the devil is in the details. Closer inspection of release trends, whether broken down by state, industrial sector, amount of releases occurring, or in connection with risk levels, suggests significant variation that is masked when looking at the broad results. Changes in environmental performance, it turns out, vary significantly from one industry to the next, one state to the next, and even from one facility to another.

The results described may also serve as initial evidence of the fact that governments and facilities need not always find themselves in the performance dilemma delineated in chapter 2. Green facilities may be reducing releases and risk simply because of their drive to meet regulatory requirements, but the dramatic decreases at a large number of facilities would suggest otherwise. There are also the interesting cases where risk levels or release levels are going in opposite directions. These categories of cases (which we refer to as blue facilities, yellow facilities, or simply "mixed" cases) also raise the possibility that governments and facilities may be motivated to escape the performance dilemma even when the payoffs are quite modest. Finally, it should be clarified that brown facilities as described in this book are not necessarily facilities failing to meet their statutory obligations. Rather, they are in most cases facilities that meet minimum expectations and not seeking to go beyond compliance.

In order to make better sense of the variation in pollution releases, analyses in the next three chapters begin to provide some answers. Chapter 4 clarifies that state-level factors are a part of the story when it comes to understanding performance variations. The TRI program was not introduced in a vacuum, but rather was implemented in states that differ from one another in their political characteristics, in their economic patterns, and in the social and cultural traits that together can affect public policy actions as well as corporate behavior. Chapters 5 and 6 complement chapter 4 by turning to detailed analyses of facility-level differences as seen through the results of our survey of facility officials. By providing a better explanation for the effects that the TRI program has had on industry and on communities, we can speak to policy changes that might be considered for the TRI or for related information provision programs. We tackle the subject of policy implications in our concluding chapter.

4

States of Green: Regional Variations in Environmental Performance

Any review of TRI data since the program's first public report in 1988 leads to an inescapable conclusion of substantial progress over time. As we recounted in chapters 1 and 3, reductions in overall releases of toxic chemicals by the nation's manufacturing industries have been truly impressive, with over 60 percent decrease in releases of the original or core chemicals tracked by the program through 2007. Yet this clear improvement in environmental performance captures only part of the full story. The national trends reflect actions taken by facilities across the fifty states, but the pattern is anything but even; there is great variation from state to state in the performance of TRI facilities. Why does this variation exist, and how does it affect conclusions that we should draw about the effects of the TRI program on industry and communities?

What is often overlooked in the reporting of national summary data is that states can vary widely in their changes from year to year. Some news reports in local papers across the country have highlighted these state-by-state differences, giving special emphasis to their own conditions. While the print media may provide some descriptive information about why changes occurred from one year to the next in their state, generally they pay little attention to the larger question of why there are state variations in the first place. Where cross-state comparisons are made, much of the time they are done at the level of comparative rankings; for example, how well or badly does a state do compared to others from one year to the next? Such variation could conceivably be random, having everything to do with particular facilities within a state and nothing to do with the states themselves. However, we have reason to believe otherwise.

For example, Maine's *Portland Press Herald* reported on a 13 percent increase in the state's 2004 pollution levels. "Maine, despite its image as

an unspoiled rural state, reported more toxic releases from industry—
9.6 million pounds—than any other New England state. It is also the
only New England state to show a year-to-year increase in releases"
(Richardson 2006). In the case of raw release levels, Maine's rising pollution can be related to polluters in the state, as all state changes can be.
While this story compared Maine to the other New England states, the
reporter did note how unique the state was. "Outside of New England,
however, Maine's toxic record looks better. Only seven states—just two
outside the region—reported fewer toxic releases in 2004 than Maine"
(Richardson 2006). Yet Maine is difficult to compare to a state with a
much larger number of industrial facilities like Ohio. The buckeye state
had nearly twenty times the pollution in 2004 as did Maine (186 million
pounds compared to 9.5 million). Unfortunately, this is the only kind of
state comparison possible with the EPA's publicly released data. How
then might one compare the policy performance of Maine and Ohio?
We chose to aggregate facilities within each state by their environmental
performance. We then compare the variation across the states in the
percentage of facilities that reduce or increase pollution emissions and
risks.

In this chapter and the following two chapters, we examine some of
the reasons for state variations in environmental performance among
facilities and states. Here we emphasize the variability in environmental
performance across the fifty states and among the thousands of companies that report through the TRI program. The companies (and the states
in which they are located) range widely in their performance over time.
Some merit the "green" label while others are clearly "brown," or
showing little or no improvement in performance, and a state's share of
either can vary dramatically. In their discussion of the "greening of
industry" trends in the mid-1990s, Press and Mazmanian observed that
company performance was either "shallow and short-term" or "deeper
and long-term" (Press and Mazmanian 1997, 255). More than ten years
later, their assessment of progress toward greener manufacturing
remained much the same (Press and Mazmanian 2010). They also began
to explain why facilities might voluntarily reduce more pollution then
they have to or, more likely, not go beyond compliance with regulatory
requirements.

With a similar bent, scholars within the field of comparative state
environmental governance describe an irregular trend toward leading
and trailing efforts by the states; that is, there is considerable variation
among the states from leaders to laggards no matter whether the actions

under investigation are state hazardous waste regulations (Daley and Garand 2005; Sapat 2004), competitive environmental federalism pressures (Potoski 2001; Woods 2006), environmental justice efforts (Ringquist and Clark 2002), toxic chemical emission trends (Grant and Jones 2004; Shapiro 2005), state requirements for green buildings (May and Koski 2007), or state natural resource innovations (Koontz 2002). A few studies have even examined the impact of additional state efforts to disseminate federal TRI information (Grant and Jones 2004; O'Toole et al. 1997; Yu et al. 1998).

States are now recognized as critical, even ascendant, regulators in the U.S. system of federalism, particularly for environmental protection policy. As Teske observed, even "in a globalized, twenty-first century, [state] regulation of industries . . . remains an important feature of the U.S. political economy" (Teske 2004, 193). Referring to these trends in his comprehensive review of state environmental policy and politics, Rabe described how scholars have concluded "that certain states tend to take the lead in most areas of policy innovation" while others consistently trail (Rabe 2010, 31). Others have observed that trailing states are in a "race to the bottom" and they lower environmental standards to attract economic development (Konisky 2007; Potoski 2001; Rabe 2010; Woods 2006).

Consider some typical findings. Potoski and Prakash (2004) found that nationwide about 6 percent of the nation's industrial facilities had voluntarily adopted the environmental management requirements of the International Organization for Standardization, or ISO 14001. When they break out adoptions by state, industrial efforts range from highs of over 20 percent in some states to less than 3 percent in others. That states vary in the production or reduction of pollution and in the management of their industrial facilities is not really in doubt. What has been missing in these kinds of studies to date is the combination of theory and data on industrial environmental performance in order to better characterize what is occurring within the states. In this chapter, we use a comparative state lens to explain why industrial environmental performance varies because of differences in political, administrative, and policy factors across the diverse system of environmental federalism. The state scale also allows a closer examination of the influence of information disclosure programs on industrial environmental performance because many states went beyond the federal TRI program with additional efforts to disseminate data on toxic chemical releases to the public.

Comparative State Environmental Performance

The performance dilemma framework from chapter 2 portrays the dynamics between two actors, a regulator and a facility. In this model, the incentives created by pollution regulation lead most facilities to compliance and most regulators to coercion. Yet, as chapter 3 showed, many facilities performed much better than others in reducing releases and risks. The regulation or performance dilemma is, as most academic models are, an oversimplification of reality. As Scholz (1991) described it, the dilemma faced by the regulators and the regulated are "nested" or embedded in a larger dynamic of political conditions that include interest groups, legislatures, the executive branch, and the public. The regulation dilemmas, including safety enforcement or industrial performance, occur within a broader series of principal-agent relationships. While a facility is considered to be an agent of its principal regulating agency, regulators are the agents of legislators, political appointees in the agencies, and interest groups. In turn, these channels of representation are potential agents of democracy's ultimate principal, the citizens. To understand how different principal-agent dynamics influence industrial environmental performance, we, like Scholz (1991), turn to a state-level analysis. The states present many variations of the social, political, and policy conditions that potentially shape the decisions by actors in the industrial environmental performance dilemma. Hence, an examination of these variations can help to clarify which of the conditions are most influential in shaping policy behavior and environmental performance in the states.

Interest in state variations has a rich lineage (Dye 1966; Dye and Gray 1980; Hofferbert 1966; Walker 1969), and its environmental policy subfield began growing in the 1980s (Lester and Bowman 1983; Lester et al. 1983a; Lester et al. 1983b; Goetze and Rowland 1985; Lester 1986; Davis and Lester 1987). In their comprehensive review of this kind of comparative state analysis, Gerber and Teske (2000) found three groups of state-level influences: (1) need or problem severity; (2) political institutions (legislature, governor, or bureaucracy), and (3) interest groups. For example, some studies found that pollution severity has a significant influence on state policy adoptions (Bacot and Dawes 1997; Lester et al. 1983a; Ringquist 1994), while others report mixed results (Potoski and Woods 2002) or found no such evidence (Lombard 1993; Williams and Matheny 1984). Other studies have suggested that administrative capacity is critical (Grant 1997; Lester et al. 1983a; Potoski and Woods 2002; Ringquist 1994).

Some of the most divergent state findings relate to political factors, with some research suggesting that interest groups or political parties are critical (Bacot and Dawes 1997; Jacoby and Schneider 2001; Ringquist 1994; Sigman 2003; Williams and Matheny 1984), while other research contradicts this conclusion (Davis and Feiock 1992; Lester et al. 1983a; Lombard 1993); one study suggested that politics influences some policy choices but not others (Potoski and Woods 2002). The influence of policy choices has attracted the attention of comparative state scholars as well. Ringquist (1993a) found regulatory stringency to be influential while others (Grant 1997; Yu et al. 1998) found that information disclosure programs are important to state TRI emission reductions. In sum, comparative studies suggest a multifaceted examination of the factors that may influence industrial environmental performance across the states. Gerber and Teske concluded that researchers ". . . cannot understand bureaucratic regulatory actions by focusing only on a single institution or even a specific category of determinants" (Gerber and Teske 2000, 876). Key categories of variables potentially include both political and administrative factors. Regulatory and nonregulatory variations across states may also be critical and cannot be excluded. Finally, control variables such as the severity of environmental problems or economic conditions must also be examined.

Several states have added collaborative approaches to their traditional deterrence-based regulation of industrial pollution. As one vein of collective action theory and its proponents predicted (Potoski and Prakash 2004; Scholz 1991), these more cooperative approaches to regulation corresponded to more optimal performance outcomes. A specific mix of state conditions strongly correlates with some states and their facilities moving beyond the collective action dilemmas discussed in chapter 2.

This rich literature on comparative state environmental policies and outcomes, along with the performance dilemma perspective from chapter 2, informs our analysis here. The questions we address are similar to those we posed at the end of chapter 1 and in our discussion of the scholarly literature on information disclosure and regulation in chapter 2, except that here we focus on the effect of state variations on the manufacturing facilities located within them. Why are some companies in some states more successfully achieving greener manufacturing goals than are others? What difference do the state regulatory and political conditions make? Moreover, how influential is information disclosure compared to other state factors? Does greater governmental capacity lead to better overall facility environmental performance within a state? Do states with bigger environmental budget commitments have more facilities that are reducing toxics? Does a state's liberalism correspond to better industrial environmental performance? Remarkably, we have little systematic inquiry into the factors that improve state industrial environmental performance, particularly pollution releases and risk reduction, and thus to what might enhance environmental quality and improve public health.

Describing State Industrial Environmental Performance Directions

We first describe state variations in industrial environmental performance using our framework from chapter 3. By using toxic chemical emission and risk levels, our descriptions contribute to our understanding of state environmental performance in ways that were impossible before the availability of the EPA's Risk Screening Environmental Indicators (RSEI) model. We then turn to explaining these variations with the more common statistical methods used in the comparative state environmental protection field: regression and path analysis. These methods allow us to sort through a variety of explanatory variables and identify the most influential causal factors connected to improvements in state industrial environmental performance.

In order to compare actions across the states, we aggregated facility-level characteristics for each state. To characterize state-level industrial environmental performance trends, we compare the percentage of a state's TRI facilities (that is, the facilities that report under the TRI program) that reduce emissions and decrease population exposure risks. However, we characterize a facility that reduces its pollution by 100 pounds the same as we do one that reduces its pollution by 1,000 pounds. Likewise, facilities decreasing or increasing their relative-risk by 10 or 1,000 receive the same characterization.

We chose to use directional categories instead of raw data for three reasons. First, as we noted in chapter 3, many analysts have challenged the accuracy of pound for pound release data from the TRI database (e.g., de Marchi and Hamilton 2006), but there is much less contention about whether facilities are increasing or reducing their release and risk levels (U.S. EPA 1993). Our aggregated measures allow for a conservative, more reliable measure of the share of a state's industrial facilities moving in the direction of safer and cleaner production. The overriding purpose in this analysis is to see which states foster more facilities moving toward more sustainable or green manufacturing (Press and Mazmanian 2010).

Second, our environmental performance measure captures the share, or percentage, of facilities within a given state that are moving in what might be termed the "right" or "wrong" direction. For instance, a study of hazardous waste reduction in Tennessee found that a measure of volume reduction allowed one large generator to mask improvements following the state's implementation of a pollution prevention program (Folz and Peretz 1997). Only 35 percent of the state's facilities decreased hazardous waste in the three years before pollution prevention began in Tennessee, and 55 percent recorded reductions during the next three years. Using raw pounds and risk data skewed the results toward those companies that produced the greatest quantities of hazardous releases. Such analyses are important, but substantively different from our objectives here.

Third, aggregating a state's percentage of leading, above average, below average, and lagging facilities also normalizes the industrial environmental performance measure across states with significantly different industrialization levels. Ohio, for instance, contains more than 900 TRI facilities, whereas states such as Vermont and New Mexico have fewer than 50 facilities.

We begin with a sample of TRI facilities (8,389) that reported toxic releases to the EPA in 1991, 1995, and 2000. In this chapter, we restrict our analysis to the difference in pollution and risk between 1991 and 1995.[1] We begin with 1991 because it is the first year following the enactment of the 1990 Pollution Prevention Act that required industry to report not just releases to the environment, but other internal toxic waste management activities such as recycling, treatment, and the use of waste for energy production. Data reliability was also a concern for the first several years of TRI reporting. Our analysis thus relies on data that provide a broader picture of industrial environmental performance while also omitting error-prone information from the first few years of TRI reporting. We use 1995 because it allows sufficient time to pass for facilities to make the kind of improvements necessary for pollution reductions (following the reasoning of Konar and Cohen (1997) and Shapiro (2005)), and the interval provides sufficient coverage to measure significant changes over time in facility environmental performance. We do not exceed five years because most of our state-level data are measures from the late 1980s or early 1990s.

We also restricted our analysis to a common set of chemicals, or the 1991 core chemicals, to ensure comparable facility-level toxic waste trends. The EPA added hundreds of new reportable toxics in 1995, so

Table 4.1
Trends in Pollution Risk and Release Reductions for Forty-Eight States

State	Reporting TRI Facilities 1991, 1995	Release Reducers % (rank)		Risk Reducers % (rank)		Release and Risk Reducers (Green) % (rank)	
State average	174.6	51.8		47.3		38.4	
Alabama	190	50.0	(31)	46.0	(29)	36.3	(33)
Arizona	58	53.4	(19)	46.6	(26)	37.9	(27)
Arkansas	155	45.8	(39)	41.9	(38)	35.5	(35)
California	428	60.0	(7)	53.0	(11)	48.1	(6)
Colorado	51	43.1	(42)	39.2	(43)	31.4	(43)
Connecticut	137	59.9	(8)	62.0	(4)	51.1	(4)
Delaware	29	69.0	(3)	55.2	(5)	51.7	(3)
Florida	157	39.5	(43)	40.8	(41)	31.8	(41)
Georgia	241	43.2	(41)	43.6	(34)	34.0	(38)
Idaho	12	33.3	(46)	33.3	(46)	16.7	(47)
Illinois	530	53.0	(21)	48.9	(22)	40.2	(20)
Indiana	419	50.6	(29)	41.3	(40)	35.6	(34)
Iowa	137	47.4	(36)	47.4	(25)	37.2	(29)
Kansas	101	53.5	(18)	54.5	(6)	40.6	(18)
Kentucky	183	49.7	(34)	43.2	(37)	37.2	(31)
Louisiana	148	58.1	(12)	50.7	(16)	44.6	(9)
Maine	38	71.1	(1)	65.8	(2)	52.6	(2)
Maryland	65	50.8	(27)	47.7	(23)	40.0	(21)
Massachusetts	179	64.2	(6)	53.1	(10)	47.5	(7)
Michigan	343	46.6	(37)	45.8	(30)	34.4	(37)
Minnesota	167	53.3	(20)	46.1	(28)	38.3	(25)
Mississippi	123	54.5	(17)	46.3	(27)	42.3	(13)
Missouri	206	51.9	(25)	43.7	(33)	37.4	(28)
Montana	14	64.3	(5)	64.3	(3)	42.9	(12)
Nebraska	56	39.3	(44)	33.9	(45)	23.2	(46)
Nevada	13	69.2	(2)	53.8	(8)	53.8	(1)
New Hampshire	46	58.7	(9)	43.5	(35)	39.1	(24)
New Jersey	249	56.2	(15)	47.4	(24)	41.0	(17)
New Mexico	14	28.6	(47)	35.7	(44)	28.6	(45)
New York	260	57.3	(14)	53.5	(9)	43.5	(11)
North Carolina	321	51.1	(26)	43.3	(36)	35.2	(36)
North Dakota	9	22.2	(48)	22.2	(48)	11.1	(48)
Ohio	698	55.6	(16)	51.0	(15)	41.7	(14)
Oklahoma	103	52.4	(24)	41.7	(39)	34.0	(39)
Oregon	101	46.5	(38)	40.6	(42)	31.7	(42)
Pennsylvania	524	50.8	(28)	50.0	(20)	39.9	(22)

Table 4.1
(continued)

State	Reporting TRI Facilities 1991, 1995	Release Reducers % (rank)		Risk Reducers % (rank)		Release and Risk Reducers (Green) % (rank)	
Rhode Island	48	58.3	(10)	50.0	(17)	43.8	(10)
South Carolina	215	50.2	(30)	44.7	(31)	37.2	(30)
South Dakota	23	39.1	(45)	30.4	(47)	30.4	(44)
Tennessee	265	49.8	(33)	51.3	(14)	41.5	(16)
Texas	509	58.0	(13)	51.3	(13)	44.6	(8)
Utah	43	67.4	(4)	44.2	(32)	39.5	(23)
Vermont	12	58.3	(11)	50.0	(18)	50.0	(5)
Virginia	185	48.6	(35)	51.4	(12)	41.6	(15)
Washington	122	50.0	(32)	54.1	(7)	40.2	(19)
West Virginia	70	52.9	(22)	50.0	(19)	37.1	(32)
Wisconsin	377	52.8	(23)	49.1	(21)	38.2	(26)
Wyoming	9	44.4	(40)	66.7	(1)	33.3	(40)

restricting our analysis to the 1991–1995 period also prevents our comparisons from being distorted by federal regulatory changes and instead allows us to concentrate on state-to-state variations.

Emission trends are then combined with risk changes from the RSEI model that provide facility-level performance characterizations; these characterizations are then aggregated by state for comparative analyses. In table 4.1 we show the proportion of facilities within each state that decrease releases, risks, and both (the share of green facilities).

We begin our analysis with basic descriptions of the state-level variations in industrial environmental performance, followed by more sophisticated data analysis later in the chapter. Maine led all states in release reductions, with 71.1 percent of their facilities getting cleaner from 1991 to 1995. The state also ranked second for risk reductions (65.8 percent safer) alone and the combined reductions of risk and releases (52.6 percent safer and cleaner). Four more relatively small states joined Maine with top ten rankings for our environmental performance measures. Nevada ranked first with its proportion of green facilities and Delaware ranked third, Connecticut fourth, and Vermont fifth.

Four much larger states also earned top ten rankings for their share of safer and cleaner industrial facilities. The geographically distant

Table 4.2
Descriptive Statistics for Selected Variables

State-Level Variables	Minimum	Maximum	Mean	Standard Deviation
Percent of state TRI facilities reducing risk	0.22	0.67	0.4729	0.08514
Percent of state TRI facilities reducing releases	0.22	0.71	0.5175	0.09806
Percent of state TRI facilities reducing both (greens)	0.11	0.54	0.3845	0.08221
Percent of state TR facilities reducing risk but increasing releases (blues)	0.00	0.33	0.0884	0.05363
Percent of state TRI facilities increasing risk but reducing releases (yellows)	0.00	0.28	0.1331	0.04313
Percent of state TRI facilities increasing both (browns)	0.14	0.67	0.3941	0.10100
Wealth by median household income	20,136	41,721	28,644	5,175
Pollution severity by TRI air releases in 1991 (natural log of pounds)	5.94	8.40	7.4274	0.57360
Public liberalism by ideological identification (mean for 1976-88)	−28.00	−0.20	−14.3000	7.51452
Hazardous waste program spending (1989)	0.33	7.93	2.5660	2.20687
Administrative professionalism (1988)	−1.61	2.33	−0.0444	0.91448
Environmental group strength by per capita membership	2.50	20.20	8.4437	3.61421
Industry group strength by air polluters economic value (1989)	0.11	0.75	0.3383	0.14233

Table 4.2
(continued)

State-Level Variables	Minimum	Maximum	Mean	Standard Deviation
Pollution prevention integration (1992)	0.00	6.00	1.8195	1.55274
Manufacturing activity by percent change in production worker hours (1992–1995)	−0.06	0.37	0.0749	0.08515

Source for TRI facilities reducing or increasing risk and releases: EPA's Risk Screening Environmental Indicators Simulation; source for wealth: U.S. Census, Statistical Abstract, various years; source for 1991 total state air releases: U.S. EPA; source for ideological identification: Erikson, Wright, McIver 1993; source for hazardous waste program spending: Hall and Kerr 1991; source for administrative professionalism: Ringquist and Clark 2002; source for environmental group strength: Hall and Kerr 1991; source for industry group strength: Commerce Department. Value added by air polluters refers to the percentage of a state's gross product (GSP) added by manufacturing industries most responsible for air pollution, expressed in millions of dollars, as of 1989; source for pollution prevention integration: U.S. EPA 1993; source for percent change in production worker hours: U.S. Census 1995, Annual Survey of Manufactures.

California and Massachusetts were just a shade apart (48.1 and 47.5 percent greening), with a sixth and seventh environmental performance ranking as just under half of their TRI facilities improved their environmental performance. Meanwhile, 44.6 percent of facilities in Texas and Louisiana became safer and cleaner from 1991 to 1995, earning these southern neighbors an eighth and ninth spot. Rhode Island finished in tenth place, completing the strong showing of smaller northeastern states in the top spots for industrial environmental performance.[2]

Our top ten rankings for environmental performance also illuminate the extreme diversity of industrialization across the states. Vermont and Nevada are two of the least industrial states (12 and 13 TRI facilities, respectively), while Texas (509 TRI facilities) and California (428 TRI facilities) are two of the largest manufacturing economies in the nation.

Table 4.2 displays descriptive statistics of industrial environmental performance for the lower 48 states. On average, nearly the same percentage of state facilities decreased pollution and risk (38.4 percent) compared to those that increased pollution and risk (39.4 percent). In

Table 4.3
State Percentages of Release and Risk Reducing (Green) TRI Facilities

Twelve Lowest States 34.5% or less	Below Average 34.6% to 38.7%	Above Average 38.8% to 42.6%	Top Twelve States 42.7% or more
Michigan	Minnesota	Mississippi	Nevada
Georgia	Wisconsin	Ohio	Maine
Oklahoma	Arizona	Virginia	Delaware
Wyoming	Missouri	Tennessee	Connecticut
Florida	Iowa	New Jersey	Vermont
Oregon	South Carolina	Kansas	California
Colorado	Kentucky	Washington	Massachusetts
South Dakota	West Virginia	Illinois	Texas
New Mexico	Alabama	Maryland	Louisiana
Nebraska	Indiana	Pennsylvania	Rhode Island
Idaho	Arkansas	Utah	New York
North Dakota	North Carolina	New Hampshire	Montana

contrast, states see far fewer facilities decreasing risk while increasing releases (blue facilities, mean = 8.8 percent) or increasing risk while decreasing releases (yellow facilities, mean = 13.3 percent). For the most part, then, facilities are primarily moving in a positive or negative direction across the states.

Table 4.3 displays a quartile separation of states by their relative percent of green TRI facilities, and figure 4.1 represents the same classification of states displayed on a map of the country; the figure permits a quick view of states with the most and the least proportion of green facilities. This approach divides states for illustrative purposes into four categories with a top tier of leaders, a second tier of above average states, a third tier below average, and a bottom tier of trailers.[3] The top tier included states with more than 42.7 percent of state facilities becoming safer and cleaner. The second quartile held states whose relative share of industrial environmental performers exceeded the median (38.7 percent). Three remaining Northeastern states (New Jersey, Pennsylvania, and New Hampshire) ranked among the top 25 for safer and cleaner facilities. Four southern states also ranked above the average: Maryland, Mississippi, Virginia, and Tennessee. Midwestern states in this second tier included Ohio, Kansas, and Illinois. Utah and Washington were the only western states appearing in this second tier of state industrial environmental performance.

State Percentages of Release and Risk Reducing (Green) TRI Facilities

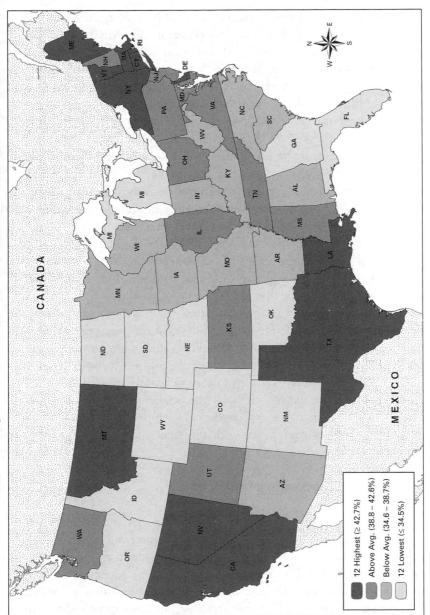

Figure 4.1
State Percentages of Release and Risk Reducing (Green) TRI Facilities.
Cartography by Stefan Freelan, Western Washington University. Data from U.S. ETA and ESRI.

The third grouping of states saw 34.6 to 38.7 percent of their facilities achieve green industrial environmental performance. Arizona was the sole western state here. The Midwest and South dominate this block of states. Minnesota, Wisconsin, Missouri, Iowa, and Indiana joined South Carolina, Kentucky, West Virginia, Alabama, Arkansas, and North Carolina, where brown facilities mostly outnumbered green facilities.

The bottom group of states is where the majority of facilities were increasing release and risk levels. In Michigan, 34 percent of the state's TRI facilities moved in the green direction, but more got brown (42 percent). The other states with significant numbers of industrial plants that increased pollution and risk included five from the West (Wyoming, Oregon, Colorado, New Mexico, and Idaho), three more from the Midwest (South Dakota, Nebraska, and North Dakota), and three from the South (Georgia, Oklahoma, and Florida). In short, a great deal of industrial environmental performance variation can be seen within states. Characterizing states into quartiles provides four categories of state industrial environmental performance that mirror our approach to facilities.

All of the northeastern states fell within the top two tiers of state industrial environmental performance while most Midwestern states fell in the bottom two tiers, but other geographical clusters are not apparent. Instead, a variety of state economic, political, and institutional factors are likely to foster facility-level innovations in pollution and risk reductions. How might we understand why this variation among the states occurs? We now turn to several statistical techniques allowing us to explain the variables across the states that influence better state industrial environmental performance.

Explaining State Environmental Protection

We now have a substantial number of studies and some pertinent theory to use in trying to understand this kind of variation from state to state. On the one hand, a diversity of studies now provides a foundation for continuing advances in the comparative state environmental literature. On the other hand, collective action theory also contributes to understanding how government and industry develop innovative policies to escape the performance dilemma. This chapter draws on these perspectives and offers three directions for moving into this field's third decade.

First, research needs to bridge the gap left by studies examining only the determinants of state environmental policies or their outcomes. The earliest comparative work on state environmental protection predominantly focused on the politics of state policy commitment variations. While subsequent studies offered methodological advances for modeling state environmental results or their policy precursors, they continued to perpetuate the divide between politics and results in the state environmental story. As Ringquist (1993b) demonstrated, comparative state environmental protection is too complex to be adequately addressed with research detaching policy politics from environmental outcomes or vice versa.

Our empirical work reported in the next section follows those few studies integrating political and policy measures with environmental outcomes (Bae, Wilcoxen, and Popp 2010; Grant 1997; Ringquist 1993b; Shapiro 2005; Woods, Konisky, and Bowman 2009). Following Lewis-Beck (1977), Ringquist (1993a, 1993b, 1995), and Woods, Konisky, and Bowman (2009), we use multiple regressions to construct a path analysis of the chain of independent variables across the states that influence industrial environmental performance. In short, this technique models industrial environmental performance as a function of political and policy variables, which in turn are a function of social and economic variables. For example, an early comparative state environmental study (Lester et al. 1983a) of hazardous waste regulation described pollution severity and economic conditions as antecedent variables, then political and administrative factors as intervening variables, and policy commitment as the dependent variable. In short, path analysis layers several regressions together to estimate the chains of influence found among many variables of sequential cause and effect.

Second, much of the previous research on comparative environmental protection relied on measures that could not fully capture variations in industrial environmental performance across the fifty states. In many cases, researchers either ignored risk or used measures such as emission levels as a proxy for risk. Measuring industrial environmental performance with only emission level trends falls short on content validity.[4] Given the state of knowledge in the 1980s and 1990s, having comparative emissions information was better than working without such data. More recent advances including development of the RSEI model, allow researchers to enhance the content validity of industrial performance measures beyond the indicators used before.

The selection of different geographic units can also distort the analysis of industrial environmental performance. For example, when Shapiro (2005) examined the relationship between state factors and risk reductions, he used geographic units of 1-kilometer by 1-kilometer square cells to aggregate risk reductions between 1988 and 1996. Bae, Wilcoxen, and Popp (2010) likewise aggregated relative state risk scores by summing county data. In both cases, the pollution risk measure was effectively disconnected from the producer: manufacturing facilities. Although an analysis of units such as zip codes or similar geographic-based cells makes some sense when analyzing environmental equity, this is less true when trying better to understand industrial environmental performance.

Third, many of the pollution measures in comparative studies cover only a small number of pollutants. Ringquist (1995), for instance, relied on changes in sulfur dioxide and nitrogen oxide levels across different states. These are unquestionably important pollutants, but they represent only a tiny fraction of toxic or hazardous chemicals released into the environment by industrial sources. By using the EPA's RSEI model, we capture trends in over 300 toxic chemical pollutants released from manufacturing sites and their relative risk to surrounding communities. We join Shapiro (2005) and Bae, Wilcoxen, and Popp (2010) in using RSEI for analysis while maintaining the link between pollution and its producer.

Modeling Variations in State Industrial Environmental Performance

Building on the comparative state environmental literature described above, we compare a more traditional dependent variable of pollution releases and a new dependent measure capturing the mix of pollutants, their variations in risk, and their sources. In particular, the model is built on the supposition that administrative, political, and policy factors all shape the interaction of government and facilities striving for improved environmental performance.

We argue that political influences work through both governmental and nongovernmental channels. In an open, pluralistic society, interest groups have multiple means to communicate their preferences to industry. Pollution practices do not exist in a political vacuum, but rather occur within particular political contexts. The extent to which a state government's policies are liberal overall is expected to influence the direction of corporate behavior and the set of expectations that corporate actors within a particular state are likely to share.

Finally, the addition of risk data will highlight variation across the states that otherwise would be missed or misunderstood. In particular, our sense is that policy innovative states are likely to be moving ahead on reducing risk at the same time that they are reducing releases. Other states, although possibly making progress on releases, are less likely to take risk into account. It cannot be overemphasized that the risk data that we are using for our new dependent variable were not readily available in the early 1990s (the time period for our analyses). States and facilities may have had similar information in their possession, but it would have been piecemeal and fragmented.

Bivariate Analysis

Our analysis proceeds in four steps. We begin by analyzing the linear relationship between pairs of variables we expect to be related to each other with bivariate correlation calculations.[5] Second, these correlation results and past empirical studies of state environmental protection then are used to construct multivariate regression models where a dependent variable's distribution is linearly related to two or more independent variables. Third, we used statistical simulations to improve the interpretation of our regression coefficients in our multivariate models (King, Tomz, and Wittenberg 2000). Finally, we combined several of our regression models using path analysis in order to identify independent variables that had direct and/or indirect effects on state industrial environmental performance.

As we noted in chapter 2 and earlier in this chapter, theory suggests that a structure of direct and indirect causation exists in the causal ordering of the determinants of state environmental commitments and policy results. Following the logic of O'Toole et al. (1997), we excluded nine states (Alaska, Hawaii, Idaho, Montana, Nevada, New Mexico, North Dakota, Vermont, and Wyoming) that had very small proportions of TRI facilities. Because of the nature of our dependent variable, states that have very few TRI facilities overall can produce substantial changes in their proportion of facilities reducing emissions and risks; a few TRI facilities improving translates into large percentages of reducers and inflates a state's industrial environmental performance.

Correlates of Industrial Environmental Performance

Bivariate correlations for the independent variables on our resource, political, policy, and performance measures yielded both expected and unexpected patterns (see table 4.4). Following much of the comparative environmental policy literature, our analysis employed a measure of the pollution problem faced by the states. Using a logarithmic transformation of total toxic chemical releases in 1991 as our measure, we expected that states with higher levels of initial pollution would, with a

Table 4.4
Selected Pearson Correlations between State Environmental Performance and Resource, Political, and Policy Variables

	1	2	3	4	5	6
1. TRI emissions in 1991	1					
2. Industry group strength	.230	1				
3. Change in production levels	.137	-.422***	1			
4. Median household income	-.213	-.145	-.402***	1		
5. Public liberalism	-.262*	-.170	-.433***	.677***	1	
6. Environmental group strength	-.450***	-.283*	-.290*	.816***	.744***	1
7. Administrative capacity	.147	-.103	-.286*	.558***	.525***	.525***
8. Toxic waste expenditures	-.382**	.102	-.504***	.569***	.601***	.734***
9. State information policy	.248	.179	-.146	.229	.067	.083
10. Pollution prevention regulation	.105	-.260	-.077	.444***	.451***	.494***
11. Release reducers percent	.019	.396**	-.419***	.374**	.172	.322**
12. Risk reducers percent	.140	.444*	-.610***	.397**	.403***	.370**
13. Release & risk reducers percent	.053	.406***	-.600***	.441***	.318**	.359**

Note: See table 4.2 for details about sources for measures.
*Statistically significant at p < 0.10.
**Statistically significant at p < 0.05.
***Statistically significant at p < 0.01.

	7	8	9	10	11	12	13
1. TRI emissions in 1991							
2. Industry group strength							
3. Change in production levels							
4. Median household income							
5. Public liberalism							
6. Environmental group strength							
7. Administrative capacity	1						
8. Toxic waste expenditures	.439***	1					
9. State information policy	.145	.068	1				
10. Pollution prevention regulation	.645***	.525***	.157	1			
11. Release reducers percent	.402***	.547***	.040	.342**	1		
12. Risk reducers percent	.408***	.548***	.190	.361**	.721***	1	
13. Release & risk reducers percent	.427***	.575***	.080	.359**	.857***	.894***	1

supportive political environment, make more of a policy effort and see more facilities reducing releases and risk than states with fewer emissions.[6]

Both political measures (public liberalism and environmental group strength) produced statistically significant but negative correlations with emission levels. More pollution was occurring in states that are more conservative. These states might have more pollution-intensive industries as well. Thus, such states with bigger industry may have a political environment that would support economic growth over environmental performance. In other words, some states might have less environmental protection to attract industry, whereas others would "race to the bottom" in environmental regulation to promote economic development (Konisky 2007; Potoski 2001; Woods 2006). Industry or government in this political environment would find little encouragement or reward to "race to the top" and move beyond compliance. From this perspective, one would expect pollution control policies and industrial environmental performance to be constrained in more conservative, economic growth-oriented conditions.

Only one of the four policy measures correlated with emission levels in a way consistent with the race to the bottom expectation. The Pearson coefficient between emission levels and state toxic waste regulation expenditures was statistically significant and negative. States with more emissions spent less on toxic chemical regulation. Correlations between the two other policy variables and all three performance measures were not consistent with a race to the bottom, however; their signs were positive and statistically insignificant.

Following Potoski and Woods (2002) and Ringquist (1993a, 1993b, 1994), we also included a measure of the state economic contribution made by industries most responsible for air pollution (as a percentage of the state's gross product). States with higher levels of economic value in their manufacturing sector may have greater industry strength both economically and politically. Therefore, this approach provided another measure to test the race-to-the-bottom (or top) thesis. The negative Pearson coefficients between this measure and our two political variables reflected the conditions that would support a race to the bottom in environmental protection. Pollution-intensive industry was more economically significant in the more conservative states that also had lower rates of environmental group membership.

However, the bivariate results between toxic chemical emissions and our policy measures did not strongly support the race-to-the-bottom perspective. No policy correlates achieved statistical significance, and while the sign for administrative capacity and pollution prevention effort was expectedly negative, the sign for toxic waste expenditures was positive. States with economically significant industry spent more on toxic waste regulation. The race-to-the-bottom and performance dilemma theses are further undermined by the statistically significant yet positive correlations between industry group strength and all three of our industrial environmental performance measures. Instead, the results could support the proposition that states and the industries they regulate have picked the lowest hanging fruit; big polluters would be the easiest places for big reductions.

We also included a dynamic measure of industrial activity measured by the change in a state's manufacturing production work hours. We intended to capture the impact on industrial environmental performance from more production (negative impact expected) or less (positive correlation expected). This correlation produced the anticipated negative and statistically significant Pearson coefficients. As manufacturing production increased, industry's environmental performance across the states declined.

Our final resource measure was median household income and, as expected, it produced a significantly positive correlation with our measures of political conditions, policy effort, and state industrial environmental performance. Wealthier states had more liberal political environments, more policy capacity and commitment, and better industrial environmental performance.

Political Determinants
One of our principal arguments is that there are political influences on state industrial environmental performance that flow through both governmental and nongovernmental channels. The extent to which a state's public and policies are liberal overall is expected to influence the direction of corporate behavior and the set of expectations that corporate actors within a particular state are likely to share. Because of the TRI program, industrial practices no longer exist in a political vacuum.

One of our two political variables was drawn from Erikson, Wright, and McIver (1993) and captured the ideological variation of a state's citizenry and the elite of the Democratic Party. The second variable measured the number of members in three major environmental groups (Sierra Club, National Wildlife Federation, and Friends of the Earth) for every thousand citizens in a state. These measures of political preference

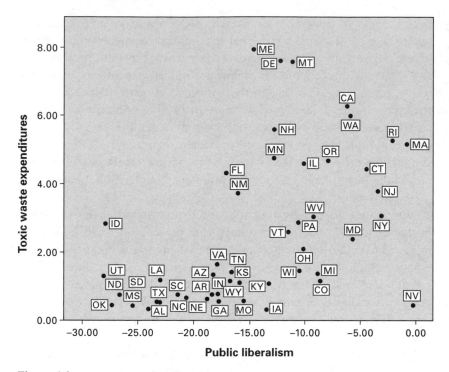

Figure 4.2
Bivariate Plot of State Liberalism and Hazardous Waste Expenditures

allow us to test the responsiveness of state policies and outcomes to democratic conditions. How responsive are regulatory bureaucracies to the political preferences of their citizens and public officials? Or, can public opinion directly, or indirectly through environmental interest groups, overcome the influence of economic interests? The statistically significant and positive correlations suggest that political preferences are at work across the states. A more liberal electorate corresponded to more protective environmental policy capacity and effort, as well as to manufactures getting safer and cleaner. States with a stronger environmental group presence also had more administrative capacity, expenditures for toxic waste control, pollution prevention, and industrial facilities that were safer and cleaner. (See figures 4.2 and 4.3.)

Policy Effort

One implication of our model is that policy commitments, although influenced by economic and political factors, are more than just a conduit

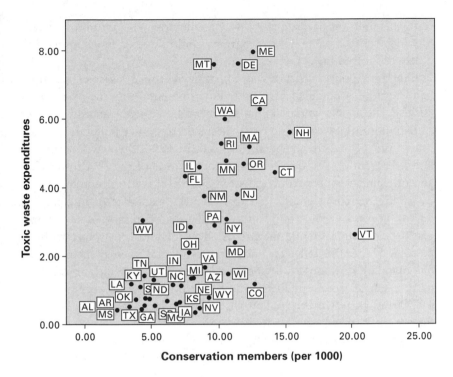

Figure 4.3
Bivariate Plot of State Conservation Group Strength and Hazardous Waste Expenditures

for these determinants. Political and economic factors affect what civil servants do, but the choices that public administrators make are by no means without discretion. Administrative autonomy is one key to variation across states. Following Ringquist and Clark (2002), we include their modified version of the administrative capacity and professionalism variable of Barrilleaux and Miller (1988) as our first policy determinants. Ringquist and Clark's institutional capacity index included state administrative salaries, training, computing power, publications, and overhead. A more professional bureaucracy corresponded with more state efforts in pollution prevention and better state industrial environmental performance.

Our second policy measure was the level of state spending on hazardous waste (per capita). The measure was obtained from Hall and Kerr's (1991) analysis of state environmental protection. Theory would suggest that states that spend more on hazardous waste should see greater improvements in their release and risk levels. The measure for per capita state spending on hazardous waste was intended to capture the extent to which states have policies, programs, and budgeted funds in place to deal with pollution.[7]

Our third policy measure was constructed from a series of questionnaire responses obtained from state environmental officials by the EPA (1993). The agency initiated an assessment of the scope and diversity of state pollution prevention activities designed to reduce pollutant releases prior to recycling, treatment, or disposal. In 1990, the new federal Pollution Prevention Act directed the administrator of the EPA to develop and implement source reduction strategies that: ". . . include the use of . . . State matching grants . . . to foster the exchange of information regarding source reduction techniques, the dissemination of such information to businesses, and the provision of technical assistance to businesses" (Sec. 6604, 181). This new environmental strategy included just one mandate for industry. Owners and operators of facilities must file a toxic chemical source reduction and recycling report if they also filed TRI forms. Otherwise, the policy was a nonregulatory assistance program for not only states, but also industry.

To create a variable reflecting state pollution prevention efforts, we followed Yu et al. (1998) in the creation of a composite measure of pollution prevention activities in state legislation, inspections, and enforcement.[8] As noted earlier, Massachusetts was an illustrative case of the leading edge of state pollution prevention efforts in the early 1990s. In 1989, the state enacted the Toxics Use Reduction Act (TURA). This state policy focuses on pollution prevention instead of control by requiring facilities to self-evaluate and plan for process improvements (O'Rourke and Lee 2004). This legislation also established the state's Toxics Use Reduction Institute (TURI) that has been providing technical assistance and grants to Massachusetts industry to reduce the use of toxics and the generation of toxic byproducts. In the EPA study used to calculate our independent measure of pollution prevention integration, Massachusetts was noted for its well-established efforts in legislation, inspections, and enforcement. In a comparison of waste reduction performance before and after state technical assistance, companies reduced 9.4 percent more toxic chemicals after the program intervention (Reibstein 2008).

The correlations between pollution prevention and facilities reducing toxic chemical releases and risks were all significant and positive. This demonstrates how, like Massachusetts, states making more of an effort on pollution prevention saw a significant portion of their industries getting safer and cleaner and this finding was very unlikely a random pattern.

We also tested a fourth policy factor that measured state information disclosure efforts. In a 1999 study, researchers developed a 10-point index of state efforts in providing toxic chemical information to the public. Based on a 1994 National Conference of State Legislatures (NCSL) survey, Yu et al. (1998) gave states a point for their affirmative response to ten NCSL questions.[9] The state scores ranged from 0 in states like Colorado and Iowa to 8 in only New Jersey and Kansas. This measure of state information disclosure achieved statistical significance with none of our industrial environmental performance variables. Unexpectedly, additional state efforts to disclose pollution release information alone did not affect a state's share of facilities reducing releases or risk.[10]

Multivariate Analysis

We used both our bivariate analysis and past research to design multiple regression models of state industrial environmental performance using Ordinary Least Squares (OLS). Since our key independent variables are interval or ratio levels of measurement, OLS is an appropriate statistical estimation technique.[11] We considered independent variables based on their enhancement of a model's adjusted R-squared.[12] We also scrutinized our regression models with a series of common diagnostics and detected no problems that could bias our estimates; diagnostics on the regression results produced no violations of the OLS assumptions of heteroscedasticity, multicollinearity, or normality.[13]

We compare three regression models of state industrial environmental performance here. We examined the same four determinants of a state's share of: (1) facilities reducing air pollution releases; (2) facilities reducing air pollution risk; and (3) facilities reducing both. The dependent measure of state industrial risk reduction had the best model performance (adjusted R-squared = 0.569, f-ratio = 14.182). The combined dependent variable of release and risk reduction was next in model performance (adjusted R-squared = 0.519, f-ratio = 11.771) while the state industrial release reduction measure had the lowest adjusted R-squared (0.388) and f-ratio (7.333). The four resource, political, and policy determinants explained a substantial amount of the state variation in all three of our dependent measures of industrial environmental performance.

As shown in table 4.5, all three models yielded a negative sign on the beta coefficient for the percent change in a state's production worker

Table 4.5
Release and Risk Reducer Regression Results on 8,369 TRI Sample

Variable	Estimated Coefficients DV: % Release Reducers	Estimated Coefficients DV: % Risk Reducers	Estimated Coefficients DV: % Release and Risk Reducers
Manufacturing activity	−0.091 (0.159)	−0.316** (0.119)	−0.340** (0.114)
Industry group strength	0.523*** (0.081)	0.467*** (0.061)	0.410*** (0.058)
Environmental group strength	0.280* (0.003)	0.244* (0.003)	0.212 (0.003)
Pollution prevention integration	0.332** (0.007)	0.338*** (0.005)	0.335** (0.005)
Adjusted r-squared	0.388	0.569	0.519
F-ratio	7.333***	14.182***	11.771***
Number of cases	41	41	41

*Statistically significant at $p < 0.10$.
**Statistically significant at $p < 0.05$.
***Statistically significant at $p < 0.01$.

hours as expected. This surrogate of manufacturing level variations produced a statistically significant relationship with the measure of industrial risk reduction and the variable combining release and risk reducing facilities, but not a significant value for just industrial release reducers. The regression coefficients for production worker hour changes show that states experiencing manufacturing growth had, while holding all other determinants equal, fewer facilities reducing pollution risk levels or reducing both releases and risk.

Also as expected, the economic value added by air polluter industries, or industry strength, in a state corresponded to more facilities getting safer and cleaner between 1991 and 1995. A positive and statistically significant coefficient for our surrogate of industry strength appeared in all three models. Environmental group strength produced a similar effect in the two models of release reduction and risk reduction, although with less impact in the latter. Finally, states with a greater effort on pollution prevention legislation, inspections, and enforcements saw a significantly greater share of facilities reducing air toxic releases and risks.

Table 4.6
Predicted Probabilities of State Industrial Environmental Performance with High and Low Determinants

Variable	DV: % Release Reducers		DV: % Risk Reducers		DV: % Release and Risk Reducers	
	Low State	Leading State	Low State	Leading State	Low State	Leading State
Manufacturing activity	0.533 (0.013)	0.522 (0.013)	0.491 (0.009)	0.456 (0.010)	0.409 (0.009)	0.376 (0.009)
Industry group strength	0.498 (0.013)	0.548 (0.011)	0.450 (0.009)	0.490 (0.008)	0.374 (0.009)	0.405 (0.008)
Environmental group strength	0.509 (0.014)	0.543 (0.013)	0.460 (0.010)	0.486 (0.009)	0.381 (0.010)	0.402 (0.009)
Pollution prevention integration	0.508 (0.013)	0.538 (0.010)	0.456 (0.009)	0.484 (0.007)	0.376 (0.009)	0.401 (0.007)

Clarified Analysis

One limitation of the results of regression modeling is that the standard outputs (*p*-values and coefficients) make practical interpretation challenging. We therefore follow King, Tomz, and Wittenberg (2000) and use our three regression models to simulate large datasets that allow predictive values to be estimated. Using the *Clarify* procedure in the statistical software STATA, we present the prediction results with varying conditions of the four determinants in table 4.6. The left column represents the first quartile of states on the listed independent variable while holding all other determinants at their mean. The right column reports predicted dependent measure values when the listed independent variable is in the highest or fourth quartile of its distribution. In other words, the left column represents the simulated average of state industrial environmental performance for states with the lowest value for each independent variable. Clarify analysis allows one to isolate the magnitude of an independent variable's influence on the dependent variable (King, Tomz, and Wittenberg 2000).

When all other independent variables are held at their mean, the value for a state's percentage of facilities reducing air pollution levels was 53.3 percent for states falling in the lowest quartile of the distribution for the variable measuring the percent change in production worker hours. Conversely, for states in the highest quartile of this determinant, the percentage of state facilities reducing air pollution releases was 52.2 percent. Thus, when all other explanatory variables are held at their means, varying a state's percent change in production worker hours from its lowest quartile value to its highest changes the predicted share of a state's release reducing facilities by only 1percent. Likewise, the difference between the leading and last states in industry group strength was 5 percent more industrial facilities decreasing air pollution releases when all other determinants were average. States at the top for environmental group strength and pollution prevention efforts meant 4 or 3 percent more facilities reducing releases respectively, holding all other variables equal.

In sum then, the variables differentiating leading and lagging states the most for release reducing facilities were industry group strength followed by environmental group strength and state pollution prevention policy. State manufacturing production had the least impact on state industrial release reduction performance. Conversely, changes in manufacturing activity had the most impact in the risk reduction model. States with the most manufacturing growth had four and one-half percent fewer facilities reducing pollution risk levels than states whose manufacturing economy had the slowest expansion.

Falling into the bottom versus the top quartile of states for industry group strength also meant a four percent difference in a state's share of pollution risk reducing facilities. States that led in environmental group strength had nearly 3 percent more facilities reducing air pollution risk while only a 2 percent difference separated states making the most or least effort on pollution prevention.

In the third model of facilities reducing both air pollution releases and risks, 3 percent more facilities reducing both air pollution releases and risk separated the top and bottom states for industry group strength or pollution prevention policy integration. Environmental group strength differences corresponded to only a 2 percent difference in facilities reducing air releases and risk.

Path Analysis
We also developed a structural model similar to that used by Ringquist (1993a,1993b,1995), Woods, Konisky, and Bowman (2009), and Hays, Elser, and Hays (1996), by working backward from our dependent variable measuring state industrial environmental performance (see figure 4.4). In the regression models described above, the independent variable betas capture only direct effects on the dependent variable. Previous

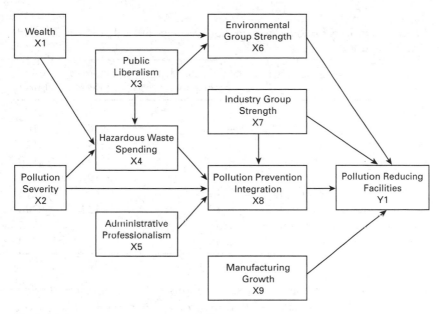

Figure 4.4
Integrated Causal Model of State Industrial Environmental Performance

comparative studies of state environmental variation established that a structure of direct and indirect causation exists between state resources, politics, policy, and environmental outcomes. Path analysis therefore was an appropriate methodology to capture both direct and indirect influences of resource, politics, and policy on state industrial environmental performance. Moreover, path analysis allowed a comparison of the combined direct and indirect effects of different variables (Lewis-Beck 1977; Ringquist 1995).

The five independent variables on the left-hand side of the structural model produced only indirect effects for our dependent measure of state industrial environmental performance. These worked indirectly through our political and policy determinants of environmental group strength and pollution prevention integration. Another indirect path flowed through two policy factors, hazardous waste spending and then pollution prevention integration. Four independent variables had direct paths to our dependent variable and only industry group strength had a direct and indirect path to pollution reducing facilities.[14]

The total effects column in table 4.7 reinforced our other results. Changes in manufacturing activity continued to be one of the most

Table 4.7
Effects Coefficients for Regression on the Percentage of Facilities Reducing Both Releases and Risk

Variable	Standardized Coefficients (Direct)	Indirect Effects	Total Effects
Wealth	—	.176	.176
Pollution severity	—	−.167	−.167
Public liberalism	—	.141	.141
Hazardous waste spending	—	.190	.190
Administrative professionalism	—	.102	.102
Environmental group strength	.212	—	.212
Industry group strength	.410	−.124	.286
Pollution prevention integration	.335	—	.335
Manufacturing production	−.340	—	−.340

influential factors influencing a state's industrial environmental performance. However, the state pollution prevention variable was a close second. Industry group strength had the largest direct effect coefficient, but the negative relation of this variable to pollution prevention decreased the total effect value. This indicated that our initial conclusions from the regression models may have overestimated the influence of this factor.

The political influence of environmental groups also appeared in the causal model both directly and as an indirect channel for state income and liberalism. Wealthier and more liberal states supported more environmentalism. States with more environmental groups saw better industrial environmental performance. Likewise, wealth and public opinion found an indirect effect path through hazardous waste spending and then pollution prevention integration. Liberal and wealthier states supported policies to foster safer and cleaner manufacturing facilities. Hazardous waste spending had the largest indirect impact on industrial environmental performance. Finally, administrative professionalism is also an indirect factor in our causal model. Better bureaucracy meant better policy and in turn, more progressive environmental performance among industrial facilities.

Discussion

Our results in this chapter reinforce those previously observed patterns of uneven state environmental performance (Lester 1994; Rabe 2010) and present a challenging puzzle to those ascribing to broad state leadership. As many of the studies in the field of comparative state environmental policy would suggest, state industrial environmental performance was far from uniform. The variations in state regulatory and political factors had a major impact on industrial environmental performance. Unexpectedly, state efforts to disclose toxic chemical information to the public had no impact on states leading in their share of pollution reducing facilities. However, integrated pollution prevention policies were a major determinant of industrial environmental performance.

In very practical terms, the differences between the eastern seaboard states of New York and Georgia mirror the uneven patterns found in our statistical modeling. They both had a similar level of industrialization in the 1990s as New York hosted 260 TRI facilities while 241 industrial plants were located in Georgia. Manufacturing grew by 7 percent in the peach state while the empire state saw industrial production worker hours shrink by 1 percent. Stark contrasts emerged when one compared the level of environmentalism, industry influence, and pollution prevention efforts in these states.

Empire state environmentalists numbered nearly 11 in 1,000 residents, and had more influence than their counterparts in Georgia, where only 5 citizens among every 1,000 were members of environmental groups. Industry accounted for 32 percent of the economic value added to the peach state economy while amounting to only 21 percent of New York's. Finally, regulators in the empire state received five out of six points for their various pollution prevention activities in the early 1990s. The state initiated legislation for both facility planning and other pollution prevention programs. Its multimedia inspections and enforcement were well developed according to the EPA. In Georgia, officials approved only facility planning legislation, partially initiated pollution prevention (P2) in their inspections, and completed some P2 permitting work, and received two out of six points for pollution prevention integration. The modeling reported here allows us to make substantive generalizations about the distance between the top and bottom states in industrial environmental performance.[15]

Our results allow us to return to the research questions introduced in chapter 2. First, we expected a greater frequency of facilities improving their environmental performance in more liberal states. The direct relation between environmental group strength and our measures of state industrial performance supports this claim. Moreover, environmental group strength was higher in states with a more ideologically liberal electorate and was an indirect factor fostering more facilities becoming safer and cleaner. Seven of the best states for industrial environmental performance also landed in the top ten of environmental group membership (NV, ME, DE, CT, VT, CA, and MA), and half of the leading twelve industrial performance states were home to the most liberal citizens (NV, CT, CA, MA, RI, and NY). The substantive influence of liberalism found here suggests that a kind of "anticipatory" industrial environmental performance occurred in the early 1990s in some states but not in others (Abel, Stephan, and Kraft 2007).

Many observers assume that pollution reductions are made voluntarily because information disclosure programs are nonregulatory in nature, but we believe that a more realistic explanation would acknowledge the incentives created by the larger political environment. In particular, we expect companies to improve pollution performance because of concerns over negative attention marshaled by environmental groups. Industries anticipated that, in the most progressive states, poor environmental performance might not go unnoticed. More specifically, state efforts to disclose pollution information to the public did not correlate with better industrial environmental performance.

Nor was it the direct impact of state program spending on hazardous waste. Rather, more per capita hazardous waste spending corresponded to more state effort to integrate pollution preventions into both regulatory and nonregulatory programs. And these policies were a key factor in those states that fostered safer and cleaner industrial facilities.

We expected better facility performance in states with greater government capacity. Again, this factor had an indirect effect on industrial environmental performance through pollution prevention integration. Only four of our top twelve states for industrial environmental performance scored in the top ten for administrative professionalism. At the top, Massachusetts led all states in this measure of government capacity. That may have been instrumental in helping the state take the lead in pollution prevention integration. However, this measure represents a state building more capacity specifically to act on the challenge of assisting industry in reducing toxic waste, instead of a generic measure of government capacity. Massachusetts is an illustrative case of the complex connections and influences somewhat masked in a quantitative model.

In 1983, Massachusetts began building its governing capacity to take on its toxic waste challenge. According to Mayer, Brown, and Linder (2002), the establishment of a pollution right-to-know law in 1983 was an achievement a full year before the Bhopal tragedy, but also did not satisfy the state's environmental activists. In particular, the Massachusetts Public Interest Research Group (MASSPIRG) built enough public support for a ballot initiative that could introduce very stringent regulation for the use of toxic substances by industry. "Fearful of this possibility," according to Mayer, Brown, and Linder (2002, 579), "industry leaders and state officials met with environmental activists and negotiated a compromise, incorporating the principles of toxics use reduction." The state's Toxics Use Reduction Act (TURA) and Toxics Use Reduction Institute (TURI) were born in 1989, and from 1991 to 1995, the state became an early leader in fostering safer and cleaner industry according to our analysis here. In short, Massachusetts was witness to the convergence of progressive environmentalists, a supportive and extremely liberal electorate, a capable state government, and a responsive industry.

Our path analysis results also represent a causal chain of principal-agent relationships (see figure 4.4 above). In the left stage of our model, citizens are the beginning principal whose expressions of political ideology relate to two agents, environmental interest groups and state government. The principals in a more liberal state citizenry find their interests pursued by stronger environmental groups and more progressive government agencies. These agents then become principals who encourage better industrial performance among the facility agents across the state.

Our analyses here produced a new representation of state environmental performance based on the frequency of facilities recording reductions in the use of air toxic emissions and their respective risk to nearby populations. The variable's structure allowed us to establish a sharp contrast between states fostering pollution reduction among their industrial plants and states where facilities increased emissions and risk. In our regression models, we showed that state industrial environmental performance was largely influenced by state interest group strength, pollution prevention policy, and manufacturing activity. These relationships were statistically significant, in the expected direction, and therefore verification of our measures' construct or convergent validity. Furthermore, a recursive path model demonstrated that state wealth, public opinion, administrative professionalism, and pollution severity indirectly influenced state environmental outcomes.

The addition of our variables that measure the frequency of facilities that reduce or increase pollution deepens our understanding of state industrial environmental performance and adds to the body of comparative state environmental research. The modeling and results here allow us to speak to several themes found in the comparative state environmental literature.

Early comparative state research concentrated on one set of determinants shaping state environmental regulation or spending. Research in the 1990s then integrated these factors into a more comprehensive perspective on state environmental performance, both with traditional policy outcomes and environmental conditions. Our empirical work here supports this integrated perspective that explains state environmental policy and outcomes as a convergence of politics, economics, and policy. States with stronger environmental groups, industry, and pollution prevention led in the greening of manufacturing in the early 1990s. States weaker in these areas trailed in the cultivation of safer and cleaner industry.

While our empirical effort supports many of the comparative environmental state studies, we also differ in an important way. This field tended to concentrate on hazardous waste, clean air, or water quality issues and not on state policies that went beyond the fragmented, media focused, or end-of-the pipe regulations dominating environmental policy over four decades. The integration that earlier state studies accomplished only encompassed the independent variables. Nor did they overcome the limitations of a much broader but influential political science perspective, principal-agent theory. It is, as Gerber and Teske (2000) noted, a simplistic or dyadic model of agency in the unfolding of state regulation. Early comparative state studies of environmental regulation were similarly narrow in the attention they gave to interest groups as the dominant principal in subnational policy efforts and in their assessment of how responsive legislative and regulating agents were to these actions. "Seeing [environmental] policymaking as a more complex environment allows multiple explanations and perhaps more room for agents to maneuver," according to Gerber and Teske (2000, 865). The performance dilemma is embedded in a larger set of principal-agent dynamics that can overcome the incentives for government to emphasize deterrence and facilities to focus on compliance only.

Some, but not all, states overcome the agency problems of information asymmetry and conflicts of interest that impede environmental performance. Like some, we found that while state capacity to lead on environmental policy is not uniform, it does vary in a systematic fashion. From another perspective, our results lend support to the perspective that state democracy is functioning well (Erikson, Wright, and McIver 1993).

Environmental policy and environmental quality tend to be better in more liberal states.

As the findings of our quantitative analysis in this chapter suggest, the resource, political, and policy conditions within states are all factors in state-by-state industrial environmental performance variations. The field of comparative state environmental studies mostly featured research trying to identify the most influential determinants of environmental leadership, spending, or even performance. These conventions and their underlying methodological commitments only to regression analysis may have fostered the empirical and theoretical fragmentation of the early comparative state environmental studies.

Yet even in those few state studies of integration, the common dependent variable also has been pollution volume. This effectively cuts off the end of the principal-agent chain, the regulated facilities producing pollution. Our dependent measure allows the completion of the principal-agent linkages from public environmentalists through the interest groups serving them and finally linking to their industrial neighbors.

Beyond the Performance Dilemma

As described before, the performance dilemma epitomizes a classic collective action problem where self-interest alone leads to individually rational choices but joint irrationality. However, the payoffs faced by government and industry do not exist in the vacuum of the idealized prisoner's dilemma. This chapter sets industrial environmental performance in a larger state context and shows that many facilities are moving toward a more optimal level of safer and cleaner manufacturing. Yet such performance was not uniform across the states in the 1990s. Instead, a specific blend of policy and political conditions was a key factor. Liberal states with dense environmental groups, robust regulations, and innovative pollution prevention policies led the way in fostering industrial environmental performance.

Our analysis also included a relatively underutilized independent policy variable in state studies, pollution prevention commitment and effort. It was employed in just two articles (O'Toole et al. 1997; Yu et al. 1998) out of the many articles on comparative state environmental studies that we reviewed earlier in this chapter. These studies were dominated by political scientists who expectedly focused on environmental policy commitments that reflected a functioning or failing state democratic system. The field also was dominated by a focus on traditional environmental policy realms like air, water, and hazardous waste. In other words, comparative state environmental studies limited their analyses to the control of pollution rather than the prevention of it.[16]

Pollution prevention activities represent one of several new strategies in a third generation of environmental history (Mazmanian and Kraft 2009). While garnering many names, such as clean production, industrial ecology, materials policy, process reengineering, and others, pollution prevention efforts share the objective of reducing waste before it is emitted into the environment. This strategy also represents a cooperative signal on the part of government that, as Potoski and Prakash described it, showed ". . . how actors in the prisoner's dilemma game can turn defection into cooperation" (Potoski and Prakash 2004, 155). Pollution prevention thus diverges from the "end-of-the-pipe" approach prevalent with past regulatory strategies and provides an important policy instrument to foster win-win cooperation and escape the performance dilemma.

However, pollution prevention policy is only a part of the industrial environmental performance story. This strategy arises out of a sequence of conditions found in some states and not others. Safer and cleaner manufacturing is a function of pollution prevention, but this policy was more likely in states with more pollution, stronger manufacturing, more hazardous waste spending, and stronger administrative capacity. These influential contexts are revealed only when the performance dilemma, principal-agent theory, and comparative state environmental perspectives are combined, as we have done here. Having said that, we also recognize how a state-level analysis offers only part of the picture. Therefore, in chapters 5 and 6, we turn to our interviews of facility and regulatory officials who are at the center of the industrial performance dilemma.

5

Facility-Level Perspectives on the TRI and Environmental Performance

The TRI has provided incentives for toxics use and waste reduction.
—Facility respondent

The [TRI] has no value to the facility. Chemical usage is tracked for operational reasons . . . independently of TRI.
—Facility respondent

The public's right to have access to TRI data has pushed industry to seek alternatives to toxic chemical use and implement source reduction activities.
—State-level public official

Not all toxic reporting is helpful or necessary.
—State-level public official

As is true of most environmental programs, progress within the TRI program is in the eye of the beholder. As the opening quotations from our respondents indicate, variation in perceptions of the program and related efforts at pollution control and prevention is common. In addition, the responses are conditioned by exactly what kind of question is asked, when it is asked, and of whom. Indeed, our surveys and interviews with both facility respondents and public officials revealed widely varying perspectives about the TRI program and the set of expectations that it represents for facility environmental performance. Some see the program as valuable and effective while others do not, and some believe that changes in the program can make it more effective while others are skeptical of what such change could bring about.

In this chapter we report on some of the distinguishing characteristics of the TRI program as seen by both facility managers and public officials. We discuss our findings about perceptions, attitudes, and behavior related to the management of toxic chemicals, drawing from both quantitative and qualitative data at our disposal. Quantitative analysis of survey results is supplemented with qualitative data from our surveys, interviews, and illustrative cases to provide a fuller picture of the views of both the regulators and the regulated. The qualitative data provide more of a "street level" assessment of how information disclosure actually works to bring about changes in corporate environmental behavior and in community decision making.

As was the case with chapter 4, our reference point is the role of mediating factors in influencing environmental performance. Figure 2.1 in chapter 2 lays out a simple analytical framework of the factors that we thought would likely influence corporate behavior in the management of toxic chemicals. As the framework indicates, we believe the environmental behavior of firms and facilities is a function of the capacity of government, facilities, communities, and the media. All influence the environmental performance at facilities around the country. Critical to our argument is the basic fact that considerable variation in "capacity" exists. As the results below will show, different types of facilities have reacted in distinct ways to the TRI program and they have responded differently when it comes to improving their environmental performance. Some factors (e.g., facility management) have had a greater role than we expected, while other factors (e.g., community groups) have had much less of a role than we expected. The results also suggest that variations over time matter as well. Many facilities have set operating procedures for dealing with the TRI program that grew out of earlier frustrations with its requirements.

Understanding the TRI Program in a Wider Context

Causal factors that drive environmental performance at the facility level are potentially numerous. As we discussed in earlier chapters, the capacity of facilities to effectively manage pollution could be driven by internal factors such as their experience with the TRI as well as external factors such as community pressure or the threat of governmental action (Fung, Graham, and Weil 2007; Grant, Jones, and Trautner 2004; Gunningham, Kagen, and Thornton 2003; Hadden 1989; Hamilton 2005; Metzenbaum 2001; Stephan 2002; Tietenberg and Wheeler 1998). Our state-level analyses presented in chapter 4 have given us an indication of a few relevant factors, but the particular context at the facility level is still unanswered. Several sources of data can provide us with a fuller understanding of the complex set of motivators at the facility and community level. In this chapter we offer an initial description of variation across facilities, including measurements of governmental, corporate, and community capacity that are central to our understanding of how mediating factors influence environmental performance. In chapter 6 we try to directly answer some of the questions about causality that we laid out in chapter 2.

To illustrate the challenge of explaining facility environmental performance, our national survey of facility TRI officials was designed to address a number of questions that cannot be answered well from the TRI dataset itself. These largely concern experience with the TRI program and the effects that it had on corporate decision making, as well as the level of interaction that facilities have with community groups. We then went a step further and surveyed a more limited sample of public officials—federal, state, and local—to get their perspectives on the TRI program and toxic chemical management more broadly. The combined survey findings allow us to better make sense of governmental, corporate, and community capacities for the management of toxic chemicals than can be determined from TRI data alone.

Survey Process

The only comparable research on the use and other effects of the TRI of which we are aware is a 1990 mail survey by Lynn and Kartez (1994), who surveyed organizations that were active users of TRI data. They identified the organizations to be studied through both written reports and referrals, with the help of EPA staff and staff affiliated with a national citizens group that monitors EPCRA implementation. State and territorial agencies were also included in the survey, as were industry groups, with the assistance of the Chemical Manufacturers Association (now the American Chemistry Council). Their survey of over 200 organizations included information on access to TRI data by individuals, their role in the dissemination of the data, their evaluations of the impact that the data had, and their views on how the TRI program might be improved. They had a high response rate of 71 percent, a function in part of dealing only with known users of TRI data and established organizations rather than individual users and formally listed corporate TRI contacts. Because we hoped to compare our findings with those of Lynn and Kartez, we employed many of the same questions related to how the TRI data might be used and the impacts the data could have.

We also sought to use the survey to help make sense of the performance dilemma that we set out in chapter 2. Do facilities find the TRI of use? What about the perceptions of the usefulness of the TRI by public officials or citizens? Is there evidence in the specific behaviors of facilities to suggest that the TRI and other factors motivate them to improve environmental performance beyond mere compliance with regulatory requirements? In addition, do public officials use the TRI as a tool in their own attempts to encourage better environmental management? The performance dilemma suggests that both facilities and public officials will be boxed into a certain corner, but a close look at what is happening on the ground can serve as a reality check for theory.

We conducted the survey between June and August of 2005, with a small number returned for several months after this period. The national survey followed a large pre-test of the survey instrument in a major metropolitan area in the Midwest in early 2005. We chose to rely on a mail questionnaire that would examine the effects of TRI experience on attitudes and behavior of three different sets of actors: corporate officials, public officials, and citizens. We sought a combination of structured questions that would permit quantitative analysis and open-ended questions that would allow respondents to speak to their experience with the program and recommendations they had for improving it. Our initial search for suitable questions and a format for the questionnaires involved a review of somewhat comparable efforts by a number of scholars studying environmental policy, attitudes, and behavior.[1]

Questionnaire development followed the procedures set out by Don Dillman (2000) for use of mail and Internet surveys through a tailored design method. The essence of this approach is to tailor the questionnaire design to the target audience and to make the questionnaires sufficiently attractive and of interest to the respondents to generate a good response rate. This was done through the selection of questions that were asked as well as in the design of the cover letter and later reminders sent to survey recipients. The response rate was of keen interest because experience in similar research efforts suggested that a rate of no more than about 20 percent from corporate actors could be expected (Andrews 2004; Delmas and Toffel 2004).[2] A methodological appendix to the chapter provides details about a pre-test of the survey instrument, the survey construction itself, and its administration.

To identify the facilities to be surveyed, we implemented both purposive and random sampling procedures. First, we identified 8,476 TRI facilities that reported releases in both 1995 and 2000.[3] This national sample included only the 1991 core chemicals to ensure consistent comparisons of facility-level toxic chemical management across the compari-

Table 5.1
Survey Sample Respondents' TRI Facility Environmental Performance (Air Emissions) Associated with Increasing or Decreasing Releases and Risk

Performance Levels	Sample	National
Green facilities decreasing pollution and/or risk	49 (20.7%)	1,447 (17.2%)
Mixed facilities with records of both decreasing	162 (68.4%)	5,860 (69.9%)
Brown facilities increasing pollution and/or risk	26 (11.0%)	1,082 (12.9%)
N	237	8,389

son period.[4] The EPA's RSEI CD-ROM provided our baseline facility-level data, including the kinds and amounts of chemicals released into air, water, or land. As discussed in chapter 3, the RSEI program provides a way to estimate the relative toxicological impacts of air releases reported in the TRI. Second, we further narrowed our sampling frame to include 3,340 facilities found in counties with at least 12 TRI sites within their boundaries. We created this cutoff because we believed that counties with very few facilities were unlikely to garner attention from external actors such as community groups, state officials, and other organizations from outside the facility. Finally, we randomly selected 1,083 facilities to receive questionnaires. The characteristics of our respondents mirrored those of the survey sample itself as well as those of the much larger database of all facilities reporting in both 1995 and 2000 (see Stephan, Kraft, and Abel 2005). Table 5.1 reaffirms the basic conclusion that the survey respondents are not significantly different from the national sample of facilities.[5]

To identify the public officials to be surveyed, we focused on three key groups: the EPA regional office with oversight of the TRI program, state officials who work closely with the program, and members of the county Local Emergency Planning Committees (LEPCs) created after the enactment of EPCRA. We initially identified 227 public officials to contact.

The results discussed below are based on responses from 237 facilities and 100 public officials nationwide, which represent a response rate of over 24.2 percent of the facilities that received the survey and a response rate of 45.5 percent of all public officials who received it.[6] Response rates for public officials varied substantially by our key categories, with 80 percent of EPA regional officials responding, 58 percent of state officials, and 38 percent of LEPC members. The lower rate for the LEPCs reflects the lack of involvement of many of those committees with the TRI reports. For surveys of these kinds the response rates are within a normal range (Baruch and Holton 2008; Delmas and Toffel 2004). We have compared our facility survey respondents with our national sample and find no response bias of any significance. Thus we believe the answers here are a good representation of the views and experience of all industrial facilities reporting TRI releases throughout the 1990s and early 2000s.

Our survey results are supplemented by our illustrative cases drawn from the pool of 3,340 facilities not already contacted for the survey.[7] The illustrative cases came from six different counties throughout the United States. The sample was purposive and included counties in the West, Midwest, Northeast, and South. We chose counties with more than twelve facilities to help ensure a sustained level of industrial activity in the regions under study and in order to maximize our chances of confirming interviews. Some of the interviews for these case studies were conducted in person and others via telephone.[8] We interviewed respondents from a variety of types of facilities, including green, brown, and mixed cases. Altogether, 39 facility respondents were interviewed. Respondents for the illustrative cases were asked questions similar to those included in the facility surveys; here, however, the format was mostly semi-structured, with abundant opportunities for probing questions and detailed discussions.

Overall Environmental Performance

Presentation of the broad performance results for our survey sample serves three purposes. First, we can compare these numbers to the results for our large population of facilities. In general there is significant overlap between the two sets. Second, we can begin to draw out the differences between leading and lagging facilities, a critical area of interest in this research. Third, we can begin to describe the internal dynamics of the performance dilemma. Chapter 4 gave us compelling evidence to suggest that states vary considerably in the performance of facilities within their boundaries, reflecting state-specific political, social, economic, and institutional differences. Here we can begin to look at the variations among individual facilities regardless of their location and illuminate some of the factors influencing their internal payoff calculus.

The 237 facilities in our survey released an average of 84,238 pounds of TRI core chemicals in 1991. On average, by 1995 that number was 58,466 pounds, and by 2000 it was 38,665 pounds. On average, risk numbers dropped in a comparable manner. These facilities were also representative of the larger universe of facilities in terms of the type of industries represented. Much like the larger group of 8,389, major industries represented in the survey sample included the chemical, rubber and plastics, and metals industries. Although some variations exist (e.g., the chemical industry makes up 23 percent of the larger sample and 32 percent of the survey sample), the differences are not statistically significant.

Similar results can be seen in responses to survey questions about whether the facility had made significant improvements over the past few years in selected areas: energy recovery, recycling, reduction in toxic releases, treatment of toxics on or off site, and source reduction. The question called for agreement or disagreement with a statement that the facility had in fact made such an improvement, using a five-point Likert scale. Over 71 percent of facilities either agreed or strongly agreed that they made improvements in recycling, and 65 percent for reduction in toxic releases. Somewhat fewer facilities agreed that progress had been made in the treatment of toxics on- or off-site or in source reduction (47 percent and 59 percent, respectively). Only 43 percent of facilities agreed that they made improvement with energy recovery. In some ways it is remarkable that such large percentages of companies are reporting progress in these areas in the past few years where the TRI data themselves do not seem to indicate gains as substantial as this. We attribute the discrepancies to differing definitions of progress and different reference time periods. In addition, the survey, of course, emphasized beliefs and perceptions as distinct from measurable changes in environmental performance.

When asked about significant environmental issues at their facility, respondents felt most strongly that air emissions were an issue, with 78 percent of respondents answering in the affirmative. A strong majority (62 percent) of facility respondents also thought that hazardous waste was an issue, with a bare majority (51 percent) believing the same about water discharges from their facility. Other issues, in descending order, included solid waste (47 percent were concerned), toxic chemical management (32 percent), odors and other nuisances (23 percent), and land use planning (5 percent). In terms of the significance of particular environmental issues, industry type mattered in some cases. For example, the primary metals industry was somewhat less likely than other industries to perceive air emissions as important. Alternatively, the rubber and plastics industry was much less likely than other industries to perceive water discharges as a critical issue. Not surprisingly, the chemicals industry is the most likely to mention that toxic chemical management is a significant issue. The variation across industries reinforces the underlying complexity of pollution emissions (Ryer 2007).

Although public officials were not directly asked about gains in environmental performance, they were asked about the seriousness of a variety of environmental problems. Their answers were in relative alignment with facility respondents. Topping the list was air pollution, which has clear connections to toxic chemical releases. In contrast, when asked about "toxic chemicals" in general, fewer public officials considered it a problem, with respondents ranking it a 25 on a 100 point "thermometer" scale. A number of other problems, including water shortages, groundwater pollution, and drinking water pollution were all ranked higher.

When splitting the public officials into our three key categories, a few differences arise. Toxic chemicals were considered to be at least as serious as other environmental problems for the state and federal officials, but somewhat of less interest to local officials whose concerns focused on hazardous waste, air pollution, water shortages, and broader issues such as economic prosperity and excessive regulation or taxes. This division in priorities is a telling comment and suggests that there is a notable difference in the perceptions of public problems by officials at varying levels of government. If these findings are representative of public officials nationwide, there is less reason for local officials to pay attention to toxic chemical management when their chief focus is on other environmental issues. For example, toxic chemicals were rated at 37 by federal and state officials (the scale is 0 to 100), but only 17 by local officials. It is important to keep in mind that local officials are a varied group, including city employees, county employees, and LEPC officials who were not themselves government employees.

Environmental performance, when brought down to the facility level, shows itself to be much more complex and nuanced than summary numbers alone can express (Fiorino 2006; Grant and Jones 2004; Grant, Jones, and Trautner 2004; Gunningham, Kagen, and Thornton 2003). For example, one of the Midwest facilities included in our illustrative cases dramatically lowered its releases from 1989 to the mid-to-late 1990s, essentially to 0 in 1997 and 2000. The company did not report in several years because of this change in releases. And yet in 2001 through 2003 its releases shot up far above the earlier levels. This example is a reminder that individual facilities are not necessarily "permanently" progressing or getting worse; rather, their performance depends on a variety of factors (Press 2007, 321). In the case of this paper industry facility, the sharp rise was easily explained. The facility started burning coal and natural gas to produce its own energy, and the releases are tied almost completely to energy generation. That is, rather than buy energy from the local utility company the facility chose to use its own fuel, with dramatic increases in releases, ten times higher than they were in 1990. As this example illustrates, facilities can exhibit substantial changes in release of TRI chemicals that have little to do directly with their manufacturing processes.

Facility Capacity: Professional Orientation of TRI Contacts

Selected facility survey responses also speak to how manufacturing industries have responded to the TRI program and what related efforts they have undertaken to improve performance. The capacity of facilities to handle environmental issues can partly be understood as an indicator of the extent to which these organizations have the capacity to push beyond the performance dilemma. Regardless of whether there are differences in levels of professionalization across green and brown facilities, it is worth knowing the extent to which professionalization occurs.

Most of the facility survey respondents were either environmental, health, and safety (EHS) staff or environmental management staff (combined the two categories accounted for 65 percent of all respondents). The rest of the respondents served in a variety of capacities for their facilities, including plant managers, human resource managers, or consultants.[9]

Facility respondents came from facilities with small numbers of employees as well as from facilities with large numbers of employees. The average number of employees was 350, but the standard deviation of 894 was much larger.[10] The large standard deviation suggests that facilities ranged from the very small to the exceptionally large, adding to other important variations across facilities. Environmental, health, and safety (EHS) staff ranged from as few as zero FTE (in cases where consultants did the TRI reporting) to over 50 FTE, although the average was just under 4 FTE and the standard deviation stood at 10.6. We rate about one-half of the facilities (46.3 percent) as having a professionalized TRI contact. These were the contacts who reported being a member of a professional organization. Although TRI reporting may have started as an amateur process on the part of facilities, clearly it has progressed over the years so that many facilities are sure to give systematic training to those within the facility who do the reporting. These are also the individuals likely to communicate with state and federal officials on environmental matters and to take advantage of any offers of EPA or state technical assistance, particularly at smaller facilities. Finally, facilities with professional environmental managers and those that invest in training are generating performance capacity.

Facility respondents have held these jobs over a wide range of years. Some have spent hardly any time in the position, while others have been working on these issues for decades at the same facilities. On average, respondents have spent just under 10 years in their current position. A majority of the respondents (58 percent) saw their work on the federal TRI as only a small percentage (by our measure, accounting for less than 20 percent) of their overall workload. Only 15 percent of the sample felt their work mostly (50 percent or more of their time) related to the TRI.

We also found wide variation in the number of weeks per year that facility officials spent compiling and submitting data to the EPA, which speaks in part to the issue of how much of a burden information disclosure requirements impose on a facility or firm. On average, the TRI reporting in any given year took facilities about 3.75 weeks to complete. There was some variation across facilities, with a number of them taking eight, nine, or more weeks to complete the necessary forms. Experience in the position is uncorrelated with the percentage of the respondent's work commitment on the TRI. More experienced respondents were neither more nor less likely to vary in the extent of their work related to the TRI. Experience in the position and the percentage of the respondent's work commitment both correlated positively with the number of weeks it takes to complete the TRI. More experienced respondents spent more time completing the TRI in a given year and, not surprisingly, respondents who had a greater work commitment on the TRI also spent more time completing the TRI.

Also worth noting is that the percentage of one's work commitment on the TRI does not correlate with the number of employees at a facility or the number of EHS staff. What this suggests is that respondents at larger facilities either (a) are not dealing with more complex reporting situations than smaller facilities, or (b) are sharing their reporting duties with other staff.

These results are subject to different interpretation. Some facilities (at least 5 percent) rely on external consultants to prepare the reports, and thus regular staff will of necessity have less involvement with the TRI program. Even many who are the facility's chief TRI manager report spending relatively little time on these duties. The point was made by at least a couple of respondents that reporting is becoming easier and less time consuming as a result of improvements in the TRI reporting software. If the burden really is declining, there are implications for proposals to alter administrative rules that are designed to reduce the frequency of reporting toxic releases. It would be reasonable to wonder whether any such reductions would be a disservice to local communities who otherwise might find TRI data useful. We make some reference to these developments below and in chapter 7.

Facility respondents are not the only ones with varying levels of experience with the TRI. This is also true of public officials at the federal, state, and local level. The officials we surveyed were not necessarily TRI experts. A full 23 percent of the full sample was either somewhat unfamiliar or very unfamiliar with the TRI. Two-thirds of those unfamiliar with the TRI program were city or county employees. Not surprisingly, all eight of the federal employees surveyed were very familiar with the program. State employees were mostly either very familiar (65 percent) or somewhat familiar (32 percent) with the program. Local officials were less familiar in general, with only 12 percent very familiar and 50 percent somewhat familiar with the program. Indeed, fully 26 percent of local officials said they were "very unfamiliar" with TRI. As noted earlier, and contrary to what one might assume, LEPCs do not regularly work with TRI data or with industry on their chemical use. Their concern seems to be focused on the potential for chemical accidents and how best to respond to such an emergency rather than on the chronic health risks to which citizens might be exposed to in the community.

These results for local officials suggest that our initial beliefs that "community" pressure might influence facility environmental performance were at least partly mistaken. The local context appears not to be a factor in the performance dilemma payoff structure. Although we cannot dismiss the role of grassroots organizations with these data, we can say that local government and entities such as LEPCs may have very little understanding about the routine chemical management and environmental performance of local facilities when it comes to toxic emissions. It is worth pointing out, given the results in chapter 4, that the way in which LEPCs approach their work may vary from state to state.

Some states may have LEPCs that stay more closely informed about toxic emissions, while other states have LEPCs that do little or nothing to move beyond the basics of emergency preparedness (Haskell 2009).

Facility Capacity: Environmental Management Systems, Specific Environmental Objectives, and Management Discretion

As Potoski and Prakash (2004) argue in their discussion of the regulation dilemma, one of the areas where firms can promote voluntary action is through the addition of environmental management systems (EMS). Having an EMS does not necessarily mean the performance dilemma has been overcome, but it signals a change in the firm's stated commitment to improving environmental conditions. We asked about facility EMSs to determine what difference, if any, they made in environmental performance.

Most of the facilities (59 percent) represented in the survey have EMSs. A much smaller percentage of facilities (19 percent) have ISO certification in environmental management. Those that are not certified overwhelmingly indicate that they are not seeking certification. Part of the explanation is the high cost of certification. During our case study interviews, a few facility officials indicated that they "had every system in place" but were not formally certified simply because of the cost involved. A fair number of facility respondents suggested that the combination of having an EMS in place along with ISO 9000 (standards for quality management systems) was satisfactory and made their pursuit of ISO 14001 (standards for environmental management) unnecessary and therefore unlikely to occur. It may also be true, as suggested by Prakash and Potoski (2006), that incentives exist for many facilities to avoid the added regulatory scrutiny that can accompany ISO certification in environmental management.

In a similar vein, we asked about specific environmental objectives that go beyond regulatory requirements. About 41 percent of facilities indicated that they had such objectives for air emissions, about 35 percent for water discharges, about 28 percent for solid waste, 37 percent for hazardous waste, 21 percent for toxic chemical management, and about 17 percent for odors, noise, or other nuisances. About 32 percent of facilities that had such objectives reported that they offered employees incentives to meet them. Putting these results another way, 68 percent of facilities that said they had specific environmental objectives of this kind provided no incentives for employees to meet them. One conclusion, therefore, is that these objectives for improved environmental performance were not taken as seriously as they might have been.

Because TRI data and our surveys focus on individual facilities and their management decisions, one question that arises is whether it makes any difference for environmental performance if a facility has a greater or lesser amount of discretion with respect to corporate headquarters when a firm has more than a single facility location. We found some variation across facilities in this regard. About 13 percent of the facilities reported being heavily dependent on corporate office preferences while others had a greater degree of discretion, either because the corporate offices had delegated responsibility to the facilities or because the facilities were small and their own offices were also the main corporate offices for the company. The discretion or lack of discretion of local facilities to control their decision-making has been shown to relate to environmental performance (Grant, Jones, and Trautner 2004). In particular, there is some evidence in the literature to suggest that facilities with remote headquarters may pollute more readily than facilities where the headquarters is on site or nearby. We save detailed comments about the relationship between local discretion and environmental performance for chapter 6, but the key point here is that there seems to be a systematic difference between green facilities and all other facilities when it comes to local discretion, and our evidence suggests that green facilities actually are more likely to have a remote headquarters than other facilities.

Impressions of the TRI Program
The facility survey was designed in part to determine whether industry views the TRI program positively or not, and whether facility experience with the TRI might be a factor affecting environmental performance. The intention was to partially answer one of our fundamental questions of chapter 2: How do facilities respond to the TRI requirements for information collection and dissemination? In answer to this question we set our work into a larger context of previous research that has tried to address the broad influences of the TRI Program (Atlas 2007; Beierle 2003; Fung and O'Rourke 2000; Lynn and Kartez 1994; Lyon and Maxwell 2004; U.S. EPA 2003). The literature has been mixed. At times research has suggested that the TRI has plausible impacts on environmental management (Fung and O'Rourke 2000; Khanna, Quimio, and Bojilova 1998; Yu et al. 1998;), while at other times there has been doubt whether the TRI has much, if any, influence on what plants do (Atlas 2007). The folk wisdom on this topic has been that corporations intensely dislike the impositions on them to comply with TRI reporting.

To our surprise, we found fairly positive or neutral attitudes toward the program itself rather than a critical perspective. Some 37 percent of facilities reported that their "overall experience" with the program was positive or highly positive in comparison to only 12 percent reporting negative or strongly negative experience. Typical of negative responses was the following quote from one survey respondent: "It seems like a lot of work to provide data to a database that has minimal value to the ones collecting the data." Over 52 percent were neutral or mixed on the point. As our case study interviews help to explain, many respondents have both positive and negative experiences with the program, with a large number reporting that it has been improving in recent years, at least in terms of the ease of reporting data. As one respondent put it, "[TRI reporting] is a struggle for someone new, but gets easier as each year passes." Most of the complaints concern redundant reporting requirements, the time and cost of compiling and reporting data, uncertainty over reporting requirements, and the potential for misinterpretation of a facility's releases (especially by the community or media). Another respondent summed it up this way: "This is a lot of work to generate information that is generally misunderstood by any[one in] the public who reviews it and provides relatively small added value to other annual reporting that is required such as for air emissions and waste."

The picture was similarly mixed for public officials, but those at the state and federal level were generally positive about the program. As one state agency respondent put it, "Community Right-to-Know has been a driving force for industry to reduce their use and releases of toxic chemicals. The public's right to have access to TRI data has pushed industry to seek alternatives to toxic chemical use and implement source reduction activities." Another federal employee saw room for improvement: "For many years EPA has ranked releases based on pounds emitted only; EPA needs to move into using the TRI data in context and provide some sort of environmental risk measurement for TRI releases, like the Risk Screening Environmental Indicator." Finally, as stated above, others, such as this state agency employee, took their evaluation a step further: "Not all toxic reporting is helpful or necessary."

Use of TRI Data

Getting a sense of how both facilities and public officials use TRI data is at the heart of this book, as we stated in chapter 2. If the TRI is not actively used by at least a significant subset of the facilities and public officials, an important question is raised about the efficacy of the program. Moreover, *how* the data are used also gives us further insight into whether the TRI actually pulls facilities and public officials out of the performance dilemma or instead serves to reinforce a command-and-control mentality. Facilities and public officials are best able to give some insights into the micro-level behaviors related to information disclosure.

When it comes to their use of the TRI, 61 percent of public officials had used the TRI to locate local environmental releases or to work on a pollution problem in their geographical area. There is striking variation when public officials are separated out into categories. One hundred percent of federal officials surveyed had used the TRI and 85 percent of state officials had as well. For local officials, the rate was much lower at 43 percent. The non-use of the TRI by LEPC members can be attributed mainly to lack of awareness of the data or program, the lack of relevance to the work of the respondents, or to problems with interpreting or easily understanding the data. Over 70 percent of local officials said they would be more likely to use TRI data if it were easier to understand the environmental and health risks posed by toxic chemicals. In chapter 7 we address this kind of change in the way TRI data are reported. In a nutshell, the TRI data would be more valuable to public officials and citizens if they pointed in a clear and understandable way to real public health risks in a particular geographic area. Databases on pounds of dozens or hundreds of chemical compounds released into the local airshed fail to meet these expectations; they do not deal directly with risks and they are not easily understandable. As we indicated in chapter 4, the state context influences the payoff structure of the performance dilemma while our survey results suggest the local context is not in play. Some simple changes to the TRI could bring local factors to bear on the performance dilemma.

Use of the TRI data varied quite a bit across public officials and facilities, although a substantial percentage of federal officials (63 percent) and state officials (38 percent) used the data to educate citizens about local pollution problems. Similarly, 75 percent of federal officials and 35 percent of state officials used the TRI data to help assist them with regulation and enforcement. Sixty-three percent of federal officials and 29 percent of state officials went as far as to say that the TRI data helped them to check facility emissions against permit records. These latter results are interesting because they suggest that although the TRI is not formally understood to be regulatory information, it has been widely used as part of the regulatory process. The results also are a reminder that the TRI program is much less a replacement for command-and-control regulation and more of a complement that serves an overlapping function. These public officials are incorporating TRI information into their regulatory routines. In addition, as we indicated in chapter 4, several states have initiated nonregulatory technical assistance programs to help facilities prevent pollution, and they use TRI data to identify which facilities might need such assistance. On a related note, 50 percent of federal officials and 27 percent of state officials used the TRI to identify needs or opportunities for P2 (pollution prevention) and source reduction, indicating that public officials can and do encourage facilities to go beyond compliance.

Although local officials used the TRI data for a large variety of reasons (see table 5.2), the numbers using the data for any given reason were quite small. Just 7 percent of local officials used the TRI data to increase their knowledge of local pollution problems, while 5 percent of local officials used the TRI data to identify needs for emergency management. Other uses were limited to even fewer local officials.

Multiple Capacities: Facility-Group Interactions

Communication between facilities and other actors touches on the full list of research questions that we presented in chapter 2. Who facilities interact with and the quantity and quality of those interactions are all indicative of how facilities respond to the TRI requirements, how the TRI data are used, the nature of mediating factors, and the differences between leaders and laggards. As we noted in chapter 2, we had an expectation of public attention serving to bring social pressure onto facilities. We therefore made sure to gather data about the interactions with community and environmental groups, the media, and local and state officials.

Facility contacts were asked about their frequency of interactions with both internal and external groups on matters of toxic releases. The results, provided in table 5.3, are telling. In particular, among the groups where there tended to be low (or no) contact were the media, legislators, local community groups, and environmental organizations. The most regular contact came with groups within the facility (employees, management), along with regulators and suppliers. These results are striking because they help us to answer the question from chapter 2 of how facilities and their surrounding communities use the toxic release information. The survey results suggest that greater use is being made internally, with much less use externally by community actors. Maybe, as

Table 5.2
Reported Use of TRI Data by Public Officials. Question: "Have you ever used TRI data as a tool to work on a pollution problem in your geographical area? If yes, how did you use the TRI data? Check all that apply." The table reports the percentages checking each of the items on the list as well as the number of respondents (in parentheses).

How TRI Data Were Used	Federal Officials	State Officials	Local Officials
Compare emissions to similar facilities	87.5 (7)	29.4 (10)	1.7 (1)
Assist with regulation and enforcement	75.0 (6)	35.3 (12)	3.5 (2)
Educate citizens about local pollution problems	62.5 (5)	38.2 (13)	1.7 (1)
Compare facility emissions over time	62.5 (5)	32.4 (11)	1.7 (1)
Check facility emissions against permit records	62.5 (5)	29.4 (10)	1.7 (1)
Compare and evaluate public environmental and health risks	62.5 (5)	20.6 (7)	3.5 (2)
Increase my knowledge of local pollution problems	50.0 (4)	32.4 (11)	6.9 (4)
Identify needs and opportunities for pollution prevention and source reduction	50.0 (4)	26.5 (9)	1.7 (1)
Prepare company profile(s)	50.0 (4)	14.7 (5)	1.7 (1)
Set local, state, and regional environmental priorities	37.5 (3)	23.5 (8)	3.5 (2)
Prepare for court litigation	25.0 (2)	5.9 (2)	1.7 (1)
Assist in citizen/industry negotiations	25.0 (2)	2.9 (1)	1.7 (1)
Contact local businesses about pollution problems	12.5 (1)	11.8 (4)	1.7 (1)
Assess the adequacy of current laws	12.5 (1)	5.9 (2)	1.7 (1)
Exert public pressure on area businesses	12.5 (1)	0.0 (0)	1.7 (1)
Identify needs for emergency management	0.0 (0)	14.7 (5)	5.2 (3)
Prepare recommendations for legislation/ regulation	0.0 (0)	14.7 (5)	1.7 (1)
Contact the media	0.0 (0)	8.8 (3)	1.7 (1)
Other	12.5 (1)	5.9 (2)	1.7 (1)

N = 8 (Federal Officials); N = 34 (State Officials); N = 58 (Local Officials).

Table 5.3

Frequency of Interaction of Facilities with Groups on Pollution Control. Question: "In the last year, about how frequently has your facility interacted with the following groups on matters of pollution control as they pertain to releases of toxic chemicals?" Numbers in the table refer to the frequency of responses. The N varies because of nonresponses to some questions.

	Daily/Weekly	Monthly	Annually	Rarely/Never
Facility employees	106	59	54	11
Corporate management	64	80	57	17
Suppliers	25	51	67	70
Customers/end users	23	35	45	98
Regulators	17	63	121	26
Trade associations	7	50	63	86
Community groups	4	25	67	119
Environmental orgs.	4	27	60	124
LEPCs	2	35	149	43
Legislators	1	7	32	165
Media	1	1	16	182

suggested by Lyon and Maxwell (2004), among others, public release of data has a greater effect on facilities trying to preempt changes in regulation and/or community pressure rather than spurring direct action by nongovernmental actors likely to push for further reductions in releases. Even interactions between facilities and local emergency planning committees (LEPCs), a community-based group where we might expect regular contact, were no more frequent than business interactions, such as between customers and facilities.

The lack of community contacts is a critical finding. A core assumption behind the design of the TRI program is that making the data available to the public will lead to public outrage and protest, and this community pressure in turn will lead to corporate changes in the management of toxic chemicals. If this were true, then we would expect to find most or at least many facilities reporting frequent interaction with community or environmental groups. Yet we do not find this at all. What might explain this finding? Perhaps more interaction was common in the early years. This would help to explain anecdotal evidence suggesting that community pressures have occurred at some facilities. This might also explain why some researchers have found evidence of media coverage in the early years of TRI reporting (Hamilton 1995; Hamilton 2005; MacLean and Orum 1992). Despite possible attention in earlier years, the results suggest that currently the information is not reaching the public for one reason or another. Possible reasons may include poor media coverage, the lack of understanding by the public about the data and its relevance to their lives, or that few community organizations take enough interest in toxic chemical management to affect environmental performance.

Perhaps there never was much interaction with the public, but the tendency of researchers has been to assume that the descriptive evidence given by EPA reports, early case studies by interest groups (MacLean and Orum 1992), and surveys (e.g., Lynn and Kartez 1994) has been true across the universe of facilities when evidence suggests it is the case for only a minority of facilities. The results parallel what we know from chapter 4 about the lack of influence of state information disclosure programs on environmental performance. The data analyses from chapter 4 and the survey results reinforce each other in this regard.

Framing the data somewhat differently, it is true that for the subset of facilities that have at least monthly contact with either community or environmental groups, there is a belief that the facility's experience with the TRI helped the facility to communicate with the local community and media, and that it helped the facility to address community concerns and complaints. There seems to be some limited evidence of the efficacy of the TRI program in facility-community interactions. In this smaller number of cases we see evidence that parallels some of the earliest research on TRI that suggested citizen action is an expected outcome of the public release of toxic pollution data (Lynn and Kartez 1994; Sarokin and Schulkin 1991).

Why such limited media attention and contact with local groups for most facilities that we surveyed? One possibility is that there is a time dynamic that is occurring. Interactions with community and environmental groups may have been more common in the late 1980s and early 1990s. The TRI reports at that time received more media attention than they did in the late 1990s or in the 2000s (Hamilton 1995). There are exceptions to this general rule, however. In December 2008, *USA Today* conducted an analysis of schools nationwide, suggesting that too many of them had high numbers of toxic releases from facilities in close proximity (Morrison and Heath 2008a). Some states and at least one U.S. Senator were so alarmed by the news that they committed to following up the *USA Today* investigation with their own analyses (Morrison and Heath 2008b). In chapter 7 we return to the schools report as part of our discussion of the numerous ways that information can be made available, and the need for realistic and understandable appraisals of chemical risks.

One other explanation for the lack of facility-community interactions may be that community members go to government officials rather than dealing with facilities directly. State and federal officials were more likely than local officials to report being contacted by citizens or community groups about toxic chemicals or pollution (83 percent versus 60 percent). Put otherwise, fully 40 percent of local officials reported no such contact over the past few years. Curiously, even when the issue was possible accidents or emergencies, the local officials were no more likely to be contacted than state and federal officials. Communication about health and environmental issues were somewhat more likely to go to the federal and state officials although there were plenty of concerns expressed across the board. Federal and state officials reported a somewhat higher level of satisfaction with the results of this communication with citizens and community groups than did local officials (80 percent were very or somewhat satisfied for the former category, and 76 percent for the latter). But the broader picture here is a general level of satisfaction with the results. Only small percentages in each category reported the results as unsatisfactory.

It is not surprising that facilities have regular contact with regulators or suppliers. As one respondent said, "Pollution control equipment is monitored on a continuous basis. We discuss toxic components of materials with suppliers several times per year. We have several visits per year from regulators." Furthermore, the fact that the occasional facility has contact with a wide range of external groups is to be expected. There is some clustering that goes on, so, for example, facilities that have contact with suppliers often are also more likely to have contact with trade associations. Similarly, facilities with less contact with legislators also had less contact with community groups. One of the more interesting assessments of the situation came from surveys collected in our pilot study: "Recently we were contacted by a community environmental group about the quantity of toxics released from our plant. To be honest, this was the first time an outside group contacted us about toxic emissions." Such statements tell us that the TRI is more likely to be digested and used within a facility rather than a facility being forced to respond to outside pressure from local communities. These results, if accurate, run distinctly contrary to the usual model of how information disclosure works. Corporate behavior is not changed merely by the release of information to the public. Rather, information can influence the relationships of businesses with each other directly. Information may be particularly salient in some communities, but less so in others. As we discuss in chapter 7,

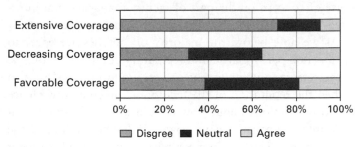

Figure 5.1
Extent of Media Coverage of Facility's TRI Releases. Question: "Please tell us the extent to which you agree or disagree with the following statements about local media coverage of the facility's TRI releases. (1) Media coverage has been extensive. (2) Media coverage has decreased over the past decade. (3) Media coverage generally has been favorable to the facility or company."

those who believe in the promise or importance of information disclosure (about pollution, or health, or school quality, or financial product safety) might want to think in terms of how best to design what information is to be released so that it is clear and meaningful to the public.

Survey data from public officials offer a slightly different lens for observing the contact between facilities and government officials. Sixty-nine percent of respondents had been contacted in the last year by at least one manufacturing facility about their releases of toxic chemicals. Despite having less experience with the TRI program, local officials were slightly more likely to have been contacted than state officials. All federal officials had been contacted in the last year, but that is not surprising given their central roles in the EPA's TRI program. The implication of these results is that contact between public officials and facilities is a two-way street. Public officials contact facilities, but the reverse is true as well.

A further set of questions dealt with media coverage of the facility's environmental performance (see figure 5.1). Just a little over 8 percent of the companies in the sample agreed that they had extensive media coverage. The majority saw little media coverage or felt the question simply did not apply to their facility. Of the small percentage that had received extensive coverage, most of them had received favorable or neutral coverage. Only a few reported extensive unfavorable coverage.

Media coverage can be understood to serve as a proxy for community concern. These results indicate very little concern existed, at least as perceived by the facilities themselves. Even so, the case study interviews suggest that media coverage does not have to be regular or extensive to stimulate a sense of concern on the part of facility managers that information reported in the TRI documents could become problematic for the company. As we note below, the majority of facilities are eager to improve environmental performance and regulatory compliance (that is, they think of these factors as important or very important). Most also seek to strengthen the reputation of the facility or the company. Hence they are likely to be cautious about how certain kinds of information may create undesired and negative news coverage that could damage the facility or firm reputation. Here again it can be argued that facilities want to preempt any undesired attention through anticipatory performance in their toxic releases (Lyon and Maxwell 2004, 68–78; Fiorino 2006, 113; Fung, Graham, and Weil 2007, 66). Facilities trending downward in their releases, or with already low levels of releases, are less likely to draw attention to themselves.

We want facilities to be accountable for their decisions, and many tell us they do want to be responsible corporate citizens. Programs like the TRI are one vehicle for helping to ensure such accountability because they force companies to be more transparent about their decisions and releases. That state and federal officials are looking at the reports means companies do have to worry about the fallout even if the local press and community groups normally would pay no attention at all. They might once a story hits the press, and that is what company officials have to worry about.

Internal factors, such as interaction with facility employees and corporate management, were cited as important. One large facility in the Midwest, for example, indicated that interaction with facility employees on pollution control of toxic chemicals occurred daily, and with corporate management and regulators weekly. The same individual reported only annual contact with LEPCs or local community groups, although he did indicate monthly interaction with environmental groups.

TRI Impacts on Facility Environmental Management

Our focus on impacts on environmental management is a corollary to the earlier issue of TRI use. Whether through direct use of TRI data or not, there is the possibility that the TRI program has had reverberations within facilities. The nature and scope of those reverberations is what interests us here.

Even though internal interactions on toxics releases were more frequent among our respondents, there were mixed views on the impact

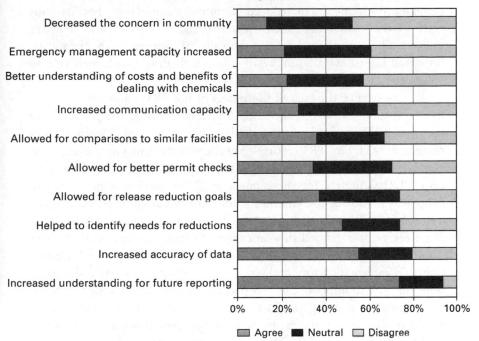

Figure 5.2
Effects of Facility Experience with the Toxics Release Inventory. Question: "Please tell us the extent to which you agree or disagree with the following statements, where 5 is 'strongly agree' and 1 is 'strongly disagree.'" Numbers in the figure refer to frequency of responses. The N varies because of nonresponses to some questions. Percentages are based on those who responded.

TRI had on environmental management decisions in their facility. As noted above, experience with the TRI program has not been as negative as many might suppose. We sought to learn more about what the TRI program actually did for the facilities. That is, in what ways did the program affect environmental management decision making?

Many facility contacts agreed that the TRI improved their understanding of reporting requirements and allowed them to gather more accurate data (see figure 5.2). Forty-eight percent further agreed that their experience with the TRI allowed them to identify needs and opportunities for source reduction, while 36 percent agreed that it allowed facilities to set goals or demonstrate commitments to release reductions. Fewer respondents agreed that the TRI requirements increased their capacity for emergency management, improved their communications with the local community and media, or reduced community complaints. Facility contacts were also mixed in their view that the TRI gave them a better understanding of the costs and benefits of dealing with chemicals, improved release data accuracy, or allowed them to check their release values against applicable permit limits.

At the extreme, the lack of usefulness of the TRI report was best put by one respondent when s/he said, "We have reached the point where each year's TRI report looks like the last. We are no longer learning anything new from our TRI report." Another respondent put it even more starkly: "We have other regulatory requirements to address these concerns. TRI is useless for these purposes." Yet there was no complete consensus in this regard. Another facility contact acknowledged that the TRI burden was minimal today: "The more I do, the easier it gets." One other respondent echoed this sentiment: "I have worked with the TRI program since it began in 1986. It has come a long way. It is a lot easier now to complete the forms and send online." Quite a few offered suggestions for how to improve the TRI program, chiefly by making it simpler and less time consuming to report information. These numbers temper the results described just above and suggest that even though a majority of facilities have not found the TRI useful for their own work, they have not as a group had a uniformly negative experience with the program.

The perspectives of public officials about the impact of the TRI were somewhat at odds with the facility responses. As table 5.4 shows, roughly 3 percent of local officials, 47 percent of state officials, and 88 percent of federal officials thought that pollution prevention actions were taken as a result of the provision of TRI information to facilities. The impact of TRI was more clearly measured by federal and state officials because of the limited ability of local officials to be of help with TRI data. Similarly, roughly 3 percent of local officials, 47 percent of state officials, and 50 percent of federal officials thought that source reduction actions were taken as a result of the provision of TRI information. These results seem to suggest that state and federal officials see a great deal of value in the TRI reports, despite some criticism about the validity of the TRI data. Officials find ways to use the data in their interactions with facilities. Having the data means that regulators have some benchmarks that they can use. TRI information can serve as leverage to motivate cooperation on the part of facilities and helps regulators to avoid confrontational policy options such as fines or orders.

Table 5.4
The Effects of Public Officials' Use of TRI Data. Question: "Please tell us whether each of the following activities, to the best of your knowledge, may have resulted from your efforts to provide information about the TRI in written reports, or by phone, mail, e-mail, personal contacts, or any other means. Check all that apply." The percentages refer to those checking each item and the number of respondents is indicated in parentheses.

Effects of Using TRI Data	Federal Officials	State Officials	Local Officials
Pollution prevention (P2) activities were undertaken	87.5% (7)	47.1% (16)	3.5% (2)
Source reduction efforts were effected at local plants	50.0 (4)	52.9 (18)	3.5 (2)
Media coverage increased	50.0 (4)	35.3 (12)	0 (0)
Industry-citizen meetings were prompted	37.5 (3)	20.6 (7)	1.7 (1)
Litigation took place or the data were used in litigation	25.0 (2)	8.8 (3)	1.7 (1)
Emergency management was improved	12.5 (1)	26.5 (9)	10.3 (6)
Legislative, regulatory, or administrative changes occurred	12.5 (1)	14.7 (5)	1.7 (1)
No activities resulted	0.0 (0)	8.8 (3)	6.9 (4)

Facility Capacity: Toxics Release Management Factors
Facility contacts were asked as well about important factors influencing their management of toxic releases. Our results suggest that corporate decision making does appear to be influenced by a variety of factors discussed previously in the literature (Coglianese and Nash 2001, 2006a; Lyon and Maxwell 2004). We also continue to answer our crucial question of facilities' response to the TRI requirements for information collection and dissemination. As suggested earlier, respondents consistently viewed legal liability and regulatory compliance as the most important while also noting that their facilities sought to achieve environmental performance improvements, save money, and strengthen their overall reputation (see figure 5.3). Respondents were divided over the importance of improving relations with environmental organizations or the community, the availability of new technologies, and deterring new regulation or legislation. Facility contacts attributed very little importance to customer loyalty, expanding business, improving employee motivation, and generating new products.

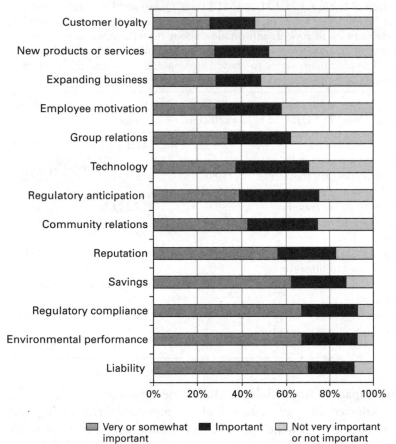

Figure 5.3
Importance of Factors in Facility Management of Toxic Chemical Releases.
Question: "For the factors listed below, please rate how important each one has
been in your facility's management of toxic chemical releases, where 5 means
'very important' and 1 means 'not at all important.'" Numbers in the figure refer
to the frequency of responses. The N varies because of nonresponses to some
questions. Percentages are based on those who responded.

Of special note, the facility sample was evenly divided over the extent to which they believed that their state supported business interests. About 37 percent agreed that the political climate in their state supported business and 41 percent disagreed with that statement. On a related note, public officials were much more likely than facility respondents to say that their state supported business interests (see chapter 7 for more details). These results hint strongly that significant variation exists among the 50 states in regulatory stringency and/or support among elected state officials (see chapter 4). Such fundamental differences in the perspectives of facilities and government officials also indicate how difficult cooperation can be, and how defection to suboptimal choices can become common. Comparative state policy research suggests that such variation helps to explain corporate behavior and perhaps even environmental quality outcomes within the states (Abel, Stephan, and Kraft 2007).

These results, combined with the impact of the TRI specifically, suggest that although many companies have not tried to improve their environmental performance because of the TRI, nevertheless, they are trying to reach a higher level of performance for other reasons (Lyon and Maxwell 2004, 22–25). Interestingly enough, despite limited contact with community-based actors, a fair number of facilities hoped to improve relations in the community. This may suggest that some facility managers believe that their image in the community can always be enhanced and that the better the relations with community-based groups, the less likelihood of future problems.

As mentioned above, simply in terms of pure behavior, over 65 percent of facilities believed they had made significant improvements over the past few years in reducing toxic releases. This suggests that reducing releases is a priority, if for reasons unrelated to the TRI reporting itself.

Other Factors

Subsequent interviews with other facilities suggested to us that the role of the TRI in helping to shape environmental policy within facilities was limited. Other requirements, including reporting and permitting requirements for air pollution or water pollution, seemed to drive their behavior more of the time. For example, one facility in Ohio mentioned that TRI reporting is only a small part of the data collection pool for the facility. Other requirements included regulatory, ISO 14001, and internal metrics.

As a facility respondent in Texas put it, "TRI is not the overwhelming drive[r]."

In the interviews, reducing costs and liability were mentioned as further motivators for better environmental performance. Both costs and liability connect back to a broader concern of companies to remain competitive with others in their industry. The ability to reduce costs and limit liability would be a critical tool for staying ahead.

There are at least a few other factors to consider when explaining environmental performance. Most facilities slightly agreed that factors such as levels of production, pollution control technology changes, and the replacement of more toxic chemicals with less toxic, had an effect on TRI releases. These examples may signify the importance of improvements in environmental performance due to factors quite removed from community pressures, media coverage, or fear of government fines. The internal drivers of the facilities may be of larger influence. Alternatively, such changes in level or process of production might themselves be a by-product of external pressures. The interviews suggest the former is more likely. When asked about which groups the respondents interacted with about pollution control, respondents generally focused on suppliers, workers, and corporate management. Community groups, environmental groups, and the media were only mentioned by a handful of facilities. Regulators were mentioned in a few cases.

There is a positive correlation between the relevance of pollution control technology changes and the frequency of contact between facilities and suppliers, trade associations, corporate management, and facility employees. These all can be understood to be either internal pressures or market-based pressures. Frequency of contact with regulators also seemed to be correlated positively with the role of pollution control technology in reducing toxic releases.

One further result is a bit puzzling. There is a negative correlation between the effect of production levels on toxic releases and the frequency of contact with either local community groups or environmental organizations. That is to say, facilities that perceive their releases to be greatly influenced by levels of production are less likely to have contact with external community actors. One possible explanation is that facilities that do fluctuate due to changes in levels of production have a ready answer for community actors worried about toxic pollution: When business is good, toxic releases go up. When business is bad, toxic releases go down.

As the next chapter will show, we found relatively few factors that are good predictors of facility performance. Many factors that are often suggested as important turn out not to be, at least with this particular dataset. How else can facilities be categorized in a way that separates facilities along other key dimensions? One answer arises from the data themselves: where facilities have specific objectives for dealing with air emissions, there also can be differences found across facilities in a number of other ways. For example, facilities with specific objectives for air pollution are more likely to talk with suppliers about pollution control than other facilities. These objective-oriented facilities also are more likely to talk with regulators, environmental organizations, and facility employees, to a name a few more.

Those facilities with specific air objectives and those without are relatively well balanced. Of the 237 facilities that answered this question, a full 41 percent of them have specific objectives, while the remaining 59 percent do not have such objectives. When placed in the context of a larger list of specific objectives, such as for water discharges, solid waste, or toxic chemical management, air objectives are the most common. In fact, 60 percent of the facilities have at least one specific objective that goes beyond regulatory requirements. And importantly, 57 percent of facilities found ways to communicate their objectives within their facility using means such as meetings, training sessions, and postings.

It is important to note that having air emissions as a significant environmental issue does not seem to be much of a driving factor. It is the *specific objectives* for air emissions where the interesting comparisons occur. Further examples abound. Facilities with specific air objectives take more time each year, on average, to deal with the TRI. These same facilities are also more likely to see changes in pollution control technology as a larger factor in their emissions decreases than other facilities.

Importantly, facilities with specific air objectives are more likely to agree that experience with the TRI allowed the facility to gather more accurate data or better estimate releases. These facilities also are more likely to agree that experience with the TRI allowed the facility to demonstrate a commitment to release reductions and resulted in fewer community complaints or expressions of concerns.

Facilities with specific air objectives have a number of incentives that drive their management of chemical releases. These facilities are more likely to be interested in customer loyalty, improving employee motivation, generating new products or services, improving regulatory compliance, and acting in advance of new legislation or regulation. They are more likely to want to improve relations with both environmental organizations and local communities.

The drivers for specific objectives for air emissions are not entirely clear, although arguably the size of emissions is part of the explanation. Large emitters are more likely to have specific air emission objectives than smaller emitters. Not surprisingly, facilities with environmental management systems are also more likely to have specific air objectives. The match is not a perfect one, but there is clearly a relationship between these two actions. Interestingly enough, facilities with specific air objectives also were more likely to perceive that their facilities received extensive media coverage. Although it is hard to know for sure, conceivably the more extensive media coverage was a partial motivator for creating specific objectives.

Facilities, Public Officials, and the Performance Dilemma

Taken as a whole, our survey results suggest mixed evidence for the ability of facilities and governments to move beyond the performance dilemma. As noted above, the main factors driving environmental management among facilities were liability and regulation. This is the compliance imperative (Fiorino 2006). However, a focus on environmental performance was a third important factor for many of the facilities. Nearly half of our respondents agreed that TRI helped identify source reduction opportunities (see figure 5.2). Similarly mixed results occurred for the public officials. A great number of federal and state officials used the TRI to assist with regulation and enforcement matters. In contrast, the responses among federal and state officials suggest that more than half of them perceived pollution prevention or source reduction being caused by the release of TRI information (see table 5.4).

Considering these results along with what we presented in chapter 4, we are beginning to get a fuller picture about the extent to which facilities vary not only in performance, but also in their reactions to pressures from the TRI program and from other factors. There is growing evidence to suggest that facilities and governments can get beyond the performance dilemma, but that many are still trapped in a traditional mindset. We now have a clearer sense of how facilities respond to the TRI requirements and a sense of how the TRI is used by both facilities and public officials (both were key research questions we posed in chapter 2). We even have some limited evidence to suggest that communities are not a significant player when it comes to use of the TRI. What we still do not know as well is what factors mediate facility behavior at the individual level. We turn to chapter 6 in order to make fuller sense of these mediating factors.

Methodological Appendix

The Survey Pretest

A preliminary corporate questionnaire was tested in late summer 2004 in a half dozen facilities located in Green Bay, Wisconsin and Portland, Oregon. These early case studies involved both personal interviews and a printed questionnaire. Our initial suspicion about the response rate was confirmed when we encountered considerable resistance by corporate officials. Some simply refused to complete the questionnaire or to schedule an interview. Others expressed concern over what kind of information we were seeking, how much of a commitment we were asking for in terms of staff time, and how the information would be used. It was apparent at this early date that corporate experience with the TRI program had been less than positive in some cases, leaving many facilities with no great desire to cooperate in such a study.

Beyond these preliminary pre-tests of the survey instrument, we decided to conduct a large pre-test to further refine the cover letters, the content of the survey, and its design. We conducted this pre-test in a large Midwestern city. For that city, we selected all of the TRI facilities that were included in our database, for which we had a TRI contact and a mailing address (86 facilities). We also identified public officials and citizen activists through a review of public records, a content analysis of newspaper coverage, and use of a reputational technique. For the pre-test we used a generous definition of public officials, which included some elected officials (for a total of 16). For the subsequent national survey we ultimately used only officials in one of three positions: the EPA regional office with oversight of the TRI program, state officials who work closely with the program, and members of the Local Emergency Management Committees created after the enactment of EPCRA. Our list of citizen activists in the pre-test city (totaling 19) also proved to be somewhat problematic because many were not as directly involved with the TRI program as we were led to believe; hence they had little interest in the survey or little to report. We concluded that for the national survey such an analysis of citizen activists would not be productive. We focused instead on the corporate and public officials surveys.

Survey Design and Administration

In light of the several pre-test results, we were careful to spell out in our initial letter of contact and in the questionnaire cover letter precisely what our objectives were, the minimal amount of time that would be necessary to complete the questionnaire (15 to 20 minutes), funding by the National Science Foundation (we thought this would help), and the sponsorship of the study by the University of Wisconsin and Washington State University.

We also revised many questions to make them as clear as possible and to frame the survey to a considerable extent to invite corporate respondents to tell us what was wrong (or right) about the TRI program and how it might be improved. We thought that would appeal to the TRI skeptics and to some extent this assumption proved to be correct. The draft questionnaires were given a more professional appearance by using Microsoft's Publisher, which allowed us to print the questionnaire in a booklet format, 7 by 8 and one-half inches.

In addition, we decided that it would be helpful to give respondents a choice of completing a paper survey or using an online version. Development of the online versions proved to be a challenge in many ways, but ultimately they offered a useful alternative that would save time in responding to the survey and would also assist us in data analysis; online responses are entered directly into an Access database. The major drawback in using the online survey, however, is that we had no e-mail addresses for any of our corporate, governmental, or citizen respondents, so they could not be given a direct link to the Web page and the survey form. Rather we had to inform them of the online alternative in our initial letters and cover letters, and in a subsequent postcard reminder, and direct them to a URL for accessing the survey. In the end, few chose the online response option. There may be some lessons here for those contemplating similar surveys in the future.

The survey itself included 28 questions for corporate respondents, seven of which involved a series of questions using a five-point Likert scale. Nearly every major question included a space for comments, and several of the questions, as noted, were open-ended (such as this one: Are there any major weaknesses of the TRI program as it applies to your facility? If so, what are those weaknesses?). Most of the questions addressed facility characteristics, experience with the TRI reporting requirements, interaction with a range of groups on pollution control issues, media coverage of the TRI, and factors that affected the facility's management of toxic releases (such as a desire to improve environmental performance, improve community relations, save money, minimize legal liability, or anticipate new legislation or regulation).

The citizen survey (used only in the pre-test) included about the same number of questions, but focused on perceptions of local environmental problems; efforts to communicate with industry, government, and others involved in pollution control; experience with the TRI program and databases; use of the TRI data in many different respects and the effects of that use; media coverage of toxics; and general questions dealing with environmental attitudes and citizen participation.

The public officials questionnaire was somewhat shorter than the other two and concentrated on perceptions of local or regional environmental problems; communication with industry and citizen groups on pollution control issues; perceptions of the regulatory environment in the community, state, or region; experience with the TRI; and the way TRI data were used and with what effects.[11]

Following Dillman's (2000) advice, we relied on a four-step approach to administration of the survey: (1) an initial letter explaining its purpose and sponsorship and indicating that the questionnaire would arrive within two weeks; (2) the survey itself, accompanied by a cover letter and a stamped, addressed return envelope; (3) a "thank-you" and reminder postcard a week later, with another statement that the survey could be accessed online with the individual's survey code number; (4) a different (and more strongly worded) cover letter and replacement questionnaire sent three weeks after the postcard. We did not use a follow-up telephone call to generate additional responses, in part because we were not convinced that the effort would make much difference.

To allow us to keep track of which respondents returned the questionnaires, we assigned a survey code number to each, which we printed on the return envelopes. The same number was to be used for the online survey. All questionnaires and letters were mailed from the University of Wisconsin-Green Bay, and all surveys were returned to the same address.

6
Environmental Leaders and Laggards: Explaining Performance

The findings presented in chapter 5 leave us with at least one clear conclusion: variation is both a hallmark of environmental performance at facilities and a key signpost of the differing opinions about the TRI program, the nature of toxic chemical management, and the extent of communication between facilities and others. Whether facilities are increasing their emissions, decreasing their emissions, or fluctuating over time, each has a unique perspective on the TRI program and management of toxic chemicals. The descriptive data from the surveys captures key aspects of the management of toxic chemical emissions, as perceived by those directly involved with the process. Yet the numerous variations across facilities still leave us with lingering questions.

Why do some companies improve their environmental performance over time while others do not? How can it be that a strong subset of industrial facilities made so little headway over the ten-year period we are examining despite the presumption in the TRI policy design that making information about toxic chemical releases public would spur a positive change in environmental decision making both within facilities and within communities? As we showed in chapter 4, there is wide variation in environmental performance over time across the nation's industrial facilities (and across the fifty states). One of the most interesting and important questions is why this is the case. How is it that federal and state environmental policies that are designed to raise the overall level of compliance and performance achieve such variable results? And what are the implications for future policy implementation or policy change that might lead to better results, that is to say, a more uniform pattern of environmental improvements over time? Our questions follow in the footsteps of much recent research on environmental performance and the greening of industry (Borck and Coglianese 2009; Coglianese and Nash 2006a; Esty and Winston 2006; Potoski and Prakash 2004; Press 2007; Press and Mazmanian 2010).

The potential of some facilities to move in dramatic fashion toward better environmental performance is well exemplified by one Midwestern facility that responded to our survey. This facility, call it the Cheese Head Company (CHC), made deep cuts in its releases between 1991 and 2000. In 1991 CHC had over 700,000 pounds of pollutants it was sending into the air. By 1995 that number had dropped to just over 85,000 pounds, and by 2001 it was down to about 5,000 pounds. The facility had a generally positive view of the TRI program, although it admitted that the program itself has had little influence on its behavior. The facility has an environmental management system in place and says that it was motivated to improve its management of toxic releases, among other reasons, because it desired to minimize legal liability, save money, and improve regulatory compliance. Clearly CHC has been an environmental leader.

Despite the national trends in TRI release and risk reductions, not all of the news has been good. While thousands of facilities have been moving in what we might call the right direction, others have been moving in the wrong direction. Some facilities, such as one from the southwestern United States, let's call it Desert Heat Incorporated (DHI), saw significant increases in releases and risk levels between 1991 and 2000. While in 1991 DHI had roughly 11,000 pounds of pollutants it was sending skyward, by 1995 that number had more than doubled to above 28,000 pounds. Just a few years later, in 2000, that number had increased almost five-fold, jumping to just under 138,500 pounds. Despite this poor environmental performance over time, DHI had a positive image of the TRI program. Like the facility mentioned at the beginning of chapter 5, DHI has an environmental management system in place and was motivated to improve its management of toxic chemical release in order to minimize any legal liability and improve regulatory compliance. Yet somehow DHI has lagged behind other facilities in its performance.

In this chapter we report facilities that have achieved the best and worst environmental performance as measured by our data on TRI releases and risk levels for the 1991 to 2000 period. We return to our analytic framework set out in chapter 2 to identify the variables that are most likely to make a difference in improving environmental performance. The data taken from the TRI for this ten-year period permit quantitative analysis of the research questions we raised in that chapter. To this we can add other data to provide a deeper understanding of what these leading facilities and firms have done and why they have been successful in comparison to others. As we noted in chapter 5, the qualitative data provide more of a "street level" assessment of how information disclosure actually works to bring about change in corporate environmental behavior and in community decision making.

Defining Environmental Performance

There are a number of ways in which to define industrial facilities as high performing or "green." Our choice, as we discussed in chapter 3, was to define green facilities as those that saw reductions in release and risk levels between 1991 and 1995 and then again between 1995 and 2000. This is not to say that *non-green* facilities are, as a group, low performers ("browns"). Many facilities have a mixed record, with some improvements, but also movement backward. As we did for green facilities, we define brown facilities based on whether they saw increasing release and risk levels between 1991 and 1995, and then again between 1995 and 2000. Non-brown facilities include both greens and the mixed cases.

Who Is Green? Distinguishing Characteristics of High Performers

Our sample of facilities with data from all three years (1991, 1995, and 2000) is quite substantial at 8,389. Of these facilities, 1,447 (17 percent) can be categorized as green. The set of facilities that we identify as green over this period of time deserve special attention because of their exceptional environmental performance. Are they different in some way from other facilities? What characteristics set them apart from other facilities? And what can this tell us about the factors that push some facilities and firms toward the green end of this spectrum of performance?

One clear difference is that green facilities began with higher levels of releases and risk levels in 1991 and end with lower levels in 2000 (see figure 6.1). Comparing the 1,447 facilities with the rest of the sample suggests that on average, there are significant differences in the amount of toxics being released and the initial risk levels in 1991. The 1,447 facilities are statistically significant larger emitters, with higher levels of risk. By 1995 the differences in releases are not statistically significant (although the 1,447 still have larger releases on average than the rest of facilities) and by 2000 the 1,447 facilities have fewer emissions on average than their counterparts and the differences are again statistically significant. The change in risk levels is even more dramatic, with the

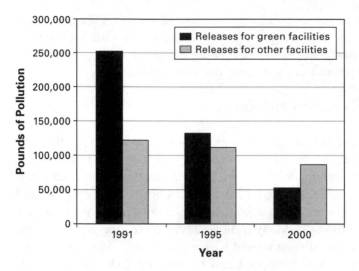

Figure 6.1
Changes in Facility Releases 1991, 1995, 2000: Greens and Others

green facilities not only having lower average risk levels by 1995, but even statistical significance by 2000 (see figure 6.2). As laid out in detail in chapter 3, risk levels (or risk scores) are complex calculations of multiple factors that relate to toxic exposures: chemical dispersion rates, facility-specific parameters such as stack height, chemical toxicity, and demographic data. Scores are theoretically limitless, although they tend not to rise above 30,000. The scores are not an exact risk assessment or risk characterization of a facility, but rather a value that can be used to compare facilities to each other in general ways.

Both figures bring into sharp relief the differences between green facilities and all others. Green facilities, as seen here, are not necessarily cleaner facilities at a given point in time, but are facilities that are making improvements in their environmental performance over time. In sum, the green facilities had more progress they needed to make compared to other facilities, but they also went farther in the ten years analyzed.

As might be expected, release and risk levels also vary among industrial sectors. The most striking results relate to the chemical industry. When compared to all other industries represented in the sample, the chemical industry is more likely to be green. Where just over 16 percent of all other facilities are likely to be green, more than 20 percent of facilities in the chemical industry are green. Not surprisingly, facilities in the chemical industry begin with higher levels of releases and risk in 1991 than other types of facilities. The chemical industry received a great deal

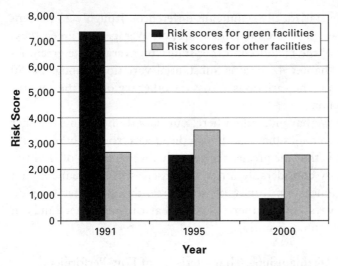

Figure 6.2
Changes in Facility Risk Scores 1991, 1995, 2000: Greens and Others. Note: Risk scores are determined by combining toxic emissions data about the amount of materials released, location of releases, toxicity of the chemicals, influence of surrounding environment, extent of human contact, and number of people affected (www.epa.gov/oppt/rsei/pubs/basic_information.html).

of attention in the late 1980s when toxic pollution and the new TRI program were more prominently featured in the nation's news. It was widely known that the industry was a major contributor to toxic releases. As we noted in chapter 3, some leading companies, such as Dow and Monsanto, gave added effort to try to reassure the public that their companies would strive to decrease their toxic pollution releases. The chemical industry also created its Responsible Care program to help promote a collective response to negative images of the industry. Many of these chemical facilities followed through on those pledges. For example, a chemical facility in California went from 164,000 pounds of toxic releases in 1991 to 125,000 pounds in 1995. By 2000, the drop was even more dramatic, as the facility's releases fell to just under 5,000 pounds of releases. A 97 percent reduction such as this is striking and speaks to the ability of at least some parts of the chemical industry to make great improvements in their environmental performance. Although other industrial sectors saw general decreases in releases and risk scores, none stood out so clearly as the chemical industry.[1]

There are some regional differences as well. Compared to all other states, those in New England (Massachusetts, New Hampshire, Vermont, Connecticut, Maine, and Rhode Island) have a greater percentage of green facilities (a statistically significant difference). Roughly 25 percent of the facilities within New England states, on average, are green. Less than 17 percent of the facilities in other states, on average, are green. As we reported in chapter 4, there is substantial variation among the 50 states, but generally the variations across the other regions of the country are relatively minor.

Beyond what we have identified here, our data do not point to other significant characteristics that distinguish the greens from all other facilities. The reason is that the greens are a highly diverse group of facilities and therefore vary considerably among themselves in the qualities that we were able to measure. It may be, of course, that green facilities do differ in other ways, and further research that can tap other kinds of data may be able to confirm this.

Who Is Brown? Distinguishing Characteristics of Low Performers

Of our sample of facilities with reported TRI data from all three years (8,389), 1,082 (or 13 percent) can be categorized as brown. They merit special attention because of their unusually weak environmental performance. Are they different in some way from other facilities? What characteristics set them apart from other facilities?

One clear difference is that consistently brown facilities began with much lower levels of emissions and risk levels in 1991 and end with higher levels in 2000 (see figure 6.3). Comparing the 1,082 facilities with the rest of the sample suggests that on average, there are significant differences in the amount of toxic chemicals being released and the initial risk levels in 1991. The 1,082 facilities are statistically significant smaller emitters with lower levels of risk. In 1995 the differences in releases remain statistically significant (the 1,082 still have lower releases on average than the rest of facilities), and by 2000 the 1,082 facilities have higher emissions on average than their counterparts (although here the differences are not statistically significant). The change in risk levels are comparable, with the consistently brown facilities having lower average risk levels in 1995 and higher average risk levels in 2000 (though not statistically significant, see figure 6.4). In sum, the consistently brown facilities not only progressed little from merely maintaining compliance, but they also regressed in environmental performance compared to other facilities in the ten years that we analyzed.

When looking at variation across industrial sectors, the most striking results relate to the plastics industry. Compared to all other industrial

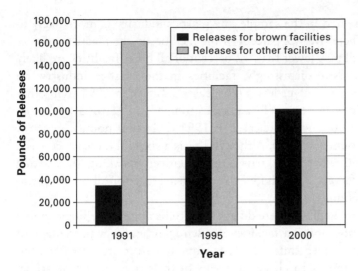

Figure 6.3
Changes in Facility Releases: 1991, 1995, 2000: Browns and Others

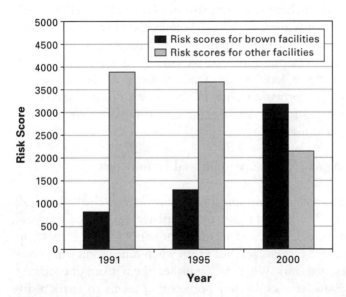

Figure 6.4
Changes in Facility Risk Scores: 1991, 1995, 2000: Browns and Others. Note: Risk Scores are determined by combining toxic emissions data about the amount of materials released, location of releases, toxicity of the chemicals, influence of surrounding environment, extent of human contact, and number of people affected (www.epa.gov/oppt/rsei/pubs/basic_information.html).

sectors represented in the sample, the plastics industry is more likely to be brown. Where over 22 percent of the facilities in the plastics industry are brown facilities, fewer than 12 percent of all other facilities are likely to be brown. Not surprisingly, facilities in the plastics industry, on average, begin with lower levels of releases and risk in 1991 than other types of facilities. For example, a plastics company in Arizona went from just over 5,000 pounds of releases in 1991 to 18,000 pounds in 1995, and 30,000 pounds in 2000. Although this is a six-fold increase between 1991 and 2000, what is also true is that, relatively speaking, the starting point for the facility is much lower than the average for all brown facilities (35,000 pounds).[2]

Unlike green facilities, there do not seem to be any compelling regional differences among brown facilities. Brown facilities, as a percentage of all facilities in a given state, vary a fair amount from state to state, but states do not appear to cluster in terms of their percentages of brown facilities.

Our analysis of the full sample is somewhat limited by our ability to gather secondary data about relevant features of the 8,389 facilities such as facility size or management practices. As more data becomes available online, there may be future opportunities to dig more deeply into a large number of facilities. For our purposes, we are able to gain some understanding from a closer look at the facilities involved in our survey dataset. Combining information from the surveys with facility-level data, we have been able to further analyze the causal factors that influence environmental performance.

Explaining Strong and Weak Environmental Performance

Variation among green, mixed, and brown facilities is clear. When all facilities are lumped together to give an overall average for a given year, much refinement is lost. The patterns of performance generate questions of causality. Why are certain facilities green when other somewhat comparable facilities are not? What distinguishes them from the others? What specific characteristics do they possess that seems to spark a different attitude toward toxic chemical management compared to other facilities and different behavior and outcomes over time that merit the green label? In chapter 2 we set out a series of questions that we are now in a position to begin to answer. We say "begin" because we cannot fully respond to the questions in light of the limitations of our data. We are not in a position to evaluate the full range of variables that may influence environmental performance. Therefore we focus on the distinctive data that we do have to address a subset of these variables.

In the pages that follow we believe we can begin to answer the following questions:

1. How does facility expertise affect environmental performance?
2. How does commitment to environmental goals by the facility (or its corporate management) affect environmental performance?
3. How does a positive facility experience affect environmental performance?
4. How does community involvement or local civic capacity affect environmental performance as perceived by facilities themselves?
5. How does local media coverage affect environmental performance as perceived by facilities themselves?
6. What other measurable factors affect environmental performance?

Answering these questions allow us to return to the analytical framework from chapter 2 and begin to see how critical factors such as corporate capacity and community capacity might influence corporate environmental decisions. Inferential analyses of the survey results help to draw connections between a variety of social, organizational, and individual factors and the decisions that facilities make. Those decisions, in turn, are expected to be a partial influence on fluctuations in environmental releases. Our results have the added value of helping us to see some limits in the analytical framework we created. As we suggest below, the role of the media and community stakeholders may be of lesser value than other key players.

Overall Environmental Performance

As we discussed in chapter 5, the 237 facilities in our survey released an average of 84,238 pounds of pollution in 1991. By 1995 that number was 58,466 pounds, and by 2000 that number was 38,665 pounds. Risk numbers dropped in a comparable manner.

Separating out green facilities from all others in the survey group is even more striking. Green facilities in the survey sample start with higher levels of releases (157,000 pounds versus 65,000 pounds), but then by 2000 the numbers are reversed (23,000 pounds for green facilities versus 43,000 pounds for all others), with green facilities having lower levels of releases. Green facilities make up about 21 percent of all facilities in the survey sample. This is only slightly higher than the roughly 17 percent of green facilities in the larger sample of sites. The changes for risk levels are quite similar. As mentioned earlier, our broad objective is to try to explain this substantial variation in environmental performance. At a minimum, however, the findings from this subgroup of facilities reinforce our suspicion that the overall improvements seen over time for the TRI program are deceptive. While releases have declined substantially since 1991, there continues to be wide variation in performance among facilities in terms of both releases of toxic chemicals and in the public health risk to surrounding communities.

The discrepancy between beliefs and measurable change are clear when comparing the reported improvements among green facilities and other facilities. Neither set of facilities differs statistically from the other when it comes to reported improvements in energy recovery, recycling, reductions in toxic releases, treatment of toxics, and source reduction. At the same time, as mentioned above, measurable improvements in releases and risk do exist and are accounted for through the categories we have created.

Separating out brown facilities in our survey sample from all others is equally as striking as when looking at green facilities. Looking at pounds of pollution in 1991 from the facilities surveyed, the evidence suggests that the browns in this group are small emitters. They are significantly smaller than other facilities; the mean for browns was around 7,600 pounds, while the mean for all others was around 94,000 pounds. By 2000, the gap is substantially lower, with the browns emitting more on average than in 1991 (23,000 pounds) and the rest of the facilities emitting less on average than in 1991 (41,000 pounds). To put it another way, between 1991 and 2000 green facilities reduced their releases an average of 85 percent while brown facilities tripled their releases, and in so doing they arrive at the same place, averaging around 23,000 pounds of pollution. Brown facilities make up only 11 percent of all facilities in the survey sample, which is slightly lower than the 13 percent for the larger sample. Changes in their risk levels parallel the descriptions above about their release levels.

Similar to green facilities, the discrepancy between beliefs and measurable change are clear when comparing the reported improvements among brown facilities and other facilities. Neither brown facilities nor all other facilities differ statistically from the other when it comes to reported improvements in energy recovery, recycling, reductions in toxic releases, treatment of toxics, and source reduction. Brown facilities were less likely to say that air emissions or water discharges were significant environmental issues for them, but the results were not very strong ($p < 0.10$).

At the same time, as noted above, measurable increases in releases and risk for brown facilities do exist and are accounted for through the categories we have created.

Facility Expertise and Environmental Performance

When comparing green facilities with all others, there are no significant differences in the professional background or experience of the TRI officials (see table 6.1). Respondents at green facilities have spent roughly the same number of years in their position as respondents at all other facilities. Respondents at green facilities were no more likely (and no less likely) than other respondents to spend the bulk of their time working on the TRI. There seems to be no relationship between the number of employees or environmental health and safety (EHS) staff and the likelihood of being categorized as a green facility. Similarly, the probability of having a professionalized TRI contact was about the same in both groups. Although it is true that on average green facilities take about three and a half weeks each year to work on TRI-related matters while non-green facilities take about five weeks, the difference is not statistically significant due to wide variances across all facilities.

Respondents at brown facilities have spent slightly more time on average in their positions as compared to all other facilities, although the results are not statistically significant. They also tend to work at facilities with smaller numbers of employees, though there is enough variation within the group of brown facilities in size to make any differences statistically nonsignificant. In all other respects mentioned above for green facilities, brown facilities are not significantly different from other facilities.

When directly comparing brown facilities and green facilities, there are some notable differences.[3] Green facilities, on average, have both larger numbers of employees overall and larger numbers of employees working on the environment, health, and safety. Aligning well with what we discussed earlier about the levels of pollutions in these facilities, there are a number of ways in which green facilities are "big" and brown facilities are "small."

Facility Environmental Management Systems and Environmental Research Performance

When asked about specific objectives that the facility might have for dealing with particular environmental issues, the only substantive differences between greens and the rest of the facilities were in the area of air

Table 6.1
Significant Differences across Facilities: Key Survey Findings

Significant Differences Between Green Facilities and All Others	Significant Differences Between Brown Facilities and All Others	Significant Differences Between Green Facilities and Brown Facilities
Less discretion at the local level for greens^ Less communication with suppliers by greens* Less Communication with Trade Associations by greens* Less communication with customers by greens* Fewer specific objectives for air emissions beyond regulatory requirements for greens* Greater decrease in media coverage for greens* Less desire to improve environmental performance by greens	Air emissions less of a major issue for browns^ Waters discharges less of a major issue for browns^ Land use planning objectives more likely for browns** Environmental Management Systems more likely to be in place for browns^ Overall satisfaction with the TRI higher for browns* Greater desire to improve environmental performance, employee motivation in environmental management*, minimize legal liability*, and strengthen their firms reputation^ for browns	More employees for greens* More EHS employees for greens* Less communication with corporate management by greens^ Less communication with trade associations by greens^ Less communication with customers by greens^ Better able to gather accurate data because of TRI experience by browns* Allowed brown facilities to check their release values against applicable permit limits* Better understanding of costs and benefits of dealing with chemicals because of TRI experience by browns^

Note: **<0.01; *<0.05; ^<0.10

Table 6.2
Difference of Means for Key Environmental Management Variables I

	Greens	All Others	t-Statistic
Discretion at local level	3.57 (0.15) N = 49	3.88 (0.08) N = 182	1.70^
Level of communication with suppliers	1.78 (0.17) N = 38	2.29 (0.09) N = 175	2.51*
Level of communication with trade associations	1.58 (0.12) N = 40	1.98 (0.07) N = 166	2.56*
Level of communication with customers	1.46 (0.15) N = 39	2.09 (0.09) N = 162	3.05**
Specific objectives for air emissions	0.27 (0.06) N = 49	0.45 (0.04) N = 188	2.32*
Decrease in media attention has occurred	3.36 (0.20) N = 28	2.88 (0.11) N = 104	–1.98*
Importance of improving environmental performance	3.60 (0.16) N = 48	3.97 (0.07) N = 178	2.29*

Note: Only key results are reported here. Full tables are available from the authors upon request. Greens = 1, Others = 0. Note: **<0.01; *<0.05; ^<0.10. Two-tailed tests. Standard errors in parentheses.

emissions (see table 6.2). Green facilities were less likely than other facilities to have specific objectives beyond regulatory requirements. This fits with the general results below; firms that are already performing well need not have specific objectives in place. If this same question had been asked of facilities in 1995 instead of 2005, the answers likely would have been very different.

The green facilities are also no more likely to have an environmental management system in place than non-green facilities. These results are interesting partly because the expectation would be that having an active EMS would play a role in influencing emissions and risk levels. The same holds true for ISO certification; however, we find that being green does not correspond to higher levels of ISO certification. Our findings are consistent with the literature that suggest a low impact of EMSs on environmental performance (Gamper-Rabindran 2006), but stand in contrast with other literature that suggests a significant impact (Andrews 2003; Eisner 2004; Potoski and Prakash 2005). As others (Potoski and Prakash 2004; Press 2007) have found, the results remain mixed on the role of EMSs on environmental performance.

Our results suggest that the existence of EMSs do correlate with other factors concerning facilities. For example, facilities with an EMS are more likely than other facilities to self-report significant improvements in environmental performance measures such as source reduction, treatment of toxics on or off site, and reduction in toxic releases. Facilities with an EMS are also more likely to report regular contact between the facility and environmental organizations, local community groups, legislators, and local emergency planning committees.[4] Not surprisingly these facilities are also more likely than their counterparts to have specific objectives in place for environmental performance that go beyond regulatory requirements as required by the EMS framework. Having specific objectives can be seen as a critical aspect of an EMS and therefore overlap is to be expected.

Facilities with an EMS are more likely than other facilities to perceive a productive purpose to the TRI. For example, experience with the TRI helped these facilities to demonstrate commitment to release reductions that was less true of facilities without an EMS. Motivation to improve a facility's management of toxic chemicals was generally higher for facilities with an EMS than other facilities. This too is not a surprise, because having an EMS itself can be seen as something that requires a significant level of motivation by the facility and/or corporate management.

There is some modest evidence to suggest that green facilities have less discretion at the local level. Although the statistical significance is weak ($p<0.10$), the data suggests that green facilities either share decision making with corporate offices more often or have limited discretion when it comes to environmental management decisions. On the one hand, since green facilities tend to be larger facilities, it would make sense that discretion is limited for them. On the other hand, a few of the interviews from our illustrative cases give some indication as to why more control from corporate offices may translate into greener facilities. It may be that in a number of cases strong leadership toward green objectives at the corporate headquarters permeated local facilities in important ways. For some, facility experience and the corporate culture meant that when the corporate office chose to go green, the local facilities simply followed through. There was no resistance and the local managers found ways to improve environmental performance, sometimes dramatically. These few interviews suggested that in certain kinds of cases, strong leadership at the corporate office was clearly the determinant factor in the facility's environmental management. Yet in other cases, the improvement in facility management was strictly a local decision. The difference is largely a function of corporate organizational structure and the prevailing culture. As we have found for other aspects of environmental

Table 6.3
Difference of Means for Key Environmental Management Variables II

	Browns	All Others	t-Statistic
Air emissions a major concern	0.65 (0.10) N = 26	0.80 (0.07) N = 211	1.73^
Water discharges a major concern	0.35 (0.10) N = 26	0.53 (0.03) N = 211	1.78^
Specific objectives for land use planning	0.12 (0.06) N = 26	0.01 (0.01) N = 211	–3.15**
EMS in place	0.75 (0.09) N = 24	0.57 (0.03) N = 206	–1.67^
Overall satisfaction with TRI	2.35 (0.11) N = 26	2.78 (0.06) N = 209	2.66**
Importance of improving environmental performance	4.24 (0.19) N = 25	3.84 (0.07) N = 201	–1.91^
Importance of improving employee motivation	3.41 (0.28) N = 22	2.73 (0.09) N = 184	–2.43*
Importance of minimizing legal liability	4.54 (0.18) N = 24	3.96 (0.08) N = 203	–2.47*
Importance of strengthening firm reputation	4.04 (0.23) N = 24	3.56 (0.08) N = 196	–1.95^
TRI experience allowed more accurate data gathering	3.91 (0.17) N = 23	3.26 (0.08) N = 205	–2.62**
TRI experience allowed check on applicable permit limits	3.48 (0.24) N = 21	2.91 (0.09) N = 199	–2.05*
TRI experience allowed better understanding of costs and benefits of dealing with chemicals	3.13 (0.25) N = 23	2.56 (0.08) N = 203	–2.22*

Note: Only key results are reported here. Full tables are available from the authors upon request. Browns = 1, Others = 0. Note: **<0.01; *<0.05; ^<0.10. Two-tailed tests. Standard errors in parentheses.

management, facilities vary greatly in the corporate organizational structure and culture, and hence in the relationship between corporate commitment and facility environmental performance. This relationship seems to be important, but further study is needed to clarify it.

When asked about specific objectives that this facility might have for dealing with particular environmental issues, the only substantive difference between browns and the rest of the facilities was in the area of land use planning (see table 6.3). Brown facilities were more likely than other facilities to have specific objectives beyond regulatory requirements.

Otherwise, we found no important distinctions when it came to specific environmental problems being addressed at the facility level.

Brown facilities are slightly more likely to have an environmental management system in place than all other facilities. If the addition of an EMS is relatively recent for brown facilities, then it might not be surprising that they are more likely to have one in place than other facilities; their continued low performance in the 1990s would serve to motivate the need for one. Green facilities may have become so either because they were early adopters of EMSs or because they were able to make progress on emissions despite having an EMS in place. It is worth keeping in mind that when just under 60 percent of all facilities have EMSs, this variable may not correlate strongly with environmental performance across facilities. Despite the result for EMSs, brown facilities are not more or less likely to be ISO certified than other facilities.

There is no evidence to suggest that brown facilities have less or more discretion at the local level when it comes to environmental management decisions. Both brown facilities and all others vary in their level of discretion.

Comparing brown and green facilities exclusively on the question of local management discretion, there are no new results not already reflected in the previous paragraphs (see table 6.4). Leaving out mixed cases may help to solidify the differences between brown and green facilities somewhat, but the overall impact of mixed cases, whether included or excluded, is relatively minor.

Although there is not a difference between green and brown facilities, it is worth noticing that facilities that have incentives for meeting specific objectives for environmental improvement are systematically different from their counterparts in a number of ways. For example, facilities that were more willing to offer incentives for meeting specific objectives were also facilities that were more likely to have an EMS in place. Similarly, these facilities were also more likely to interact regularly with both regulators and local community groups/leaders. In addition, these facilities experienced an increased ability to discuss chemical releases with the local community and media. Interestingly, these kinds of facilities were less likely to be driven by legal liability than other facilities with specific objectives. Though having specific environmental objectives is an important aspect of many facilities, providing incentives for employee environmental performance has further positive impact on environmental management. Here again, the corporate or facility management culture seems to make a significant difference in recognizing and

Table 6.4

Difference of Means for Key Environmental Management Variables III

	Greens	Browns	t-Statistic
Level of communication with corporate management	2.74 (0.16) N = 47	3.30 (0.27) N = 23	1.87^
Level of communication with trade associations	1.58 (0.12) N = 40	1.96 (0.20) N = 23	1.70^
Level of communication with customers	1.46 (0.15) N = 39	1.87 (0.17) N = 23	1.77^
Specific objectives for air emissions	0.27 (0.06) N = 49	0.50 (0.10) N = 26	2.06*
Water discharges a major concern	0.55 (0.07) N = 49	0.35 (0.10) N = 26	−1.70^
Specific objectives for land use planning	0.02 (0.02) N = 49	0.12 (0.06) N = 26	1.75^
Overall satisfaction with TRI	2.81 (0.11) N = 48	2.35 (0.11) N = 26	−2.82**
Importance of improving environmental performance	3.60 (0.16) N = 48	4.24 (0.19) N = 25	2.43*
Importance of improving employee motivation	2.59 (0.19) N = 41	3.41 (0.28) N = 22	2.52*
Importance of minimizing legal liability	3.89 (0.16) N = 47	4.54 (0.18) N = 24	2.56*
Importance of strengthening firm reputation	3.53 (0.18) N = 45	4.04 (0.23) N = 24	1.70^
Number of employees	389.3 (105.1) N = 47	99.1 (30.5) N = 26	−2.02*
Number of EHS employees	3.59 (0.63) N = 46	2.08 (0.35) N = 26	−1.71^
TRI experience allowed more accurate data gathering	3.26 (0.17) N = 46	3.91 (0.17) N = 23	2.39*
TRI experience allowed check on applicable permit limits	2.86 (0.17) N = 44	3.48 (0.24) N = 21	2.10*
TRI experience allowed better understanding of costs and benefits of dealing with chemicals	2.53 (0.18) N = 45	3.13 (0.25) N = 23	1.95^

Note: Only key results are reported here. Full tables are available from the authors upon request. Greens = 1, Browns = −1. Note: **<0.01; *<0.05; ^<0.10; two-tailed tests. Standard errors are in parentheses.

responding to environmental challenges and in improving environmental performance.

Facility Impressions of the TRI Program and Environmental Performance

Being a green facility does not seem to influence overall satisfaction with the TRI program. On average, all facilities have a slightly negative view of the TRI program, although as we noted in chapter 5, the variance is wide, with both positive and negative experiences being expressed.

Surprisingly, being a brown facility appears to correlate positively with overall satisfaction with the TRI program. Why might brown facilities have a more positive view? It is not entirely clear. Possibly brown facilities want to appear better than they are, and so declare themselves to be satisfied with the TRI. Alternatively, because brown facilities in general were lower emitters in the early 1990s, they may not have gained as much attention when the TRI results were made available publicly at that time. They could more readily feel satisfied because they were less likely to have dealt with any negative effects. That early experience may have affected their views even well into the 2000s.

Possibly more important than environmental performance was a direct sense of relevance of the TRI as perceived by facilities, whether green, brown, or mixed cases. Those facilities that believed that the TRI was affecting what they did at their facility were more likely to be satisfied with the TRI. These results make sense in a very straightforward way. Facilities that do not find TRI data useful can be expected to see this particular regulatory requirement as a waste of their limited resources. Implicit in these results are arguments for why the TRI program should be kept, but also ideas for how it could be modified and improved. The fact that some facilities make productive use of TRI information is a good sign of the program's value. Improvements could focus on figuring out what is missing from the TRI program that could make it useful to other facilities. For example, some of the facilities believed that their experience with the TRI helped them to identify needs and opportunities for source reduction. These facilities in turn were more positive about the TRI program. What would it take for other facilities to have the TRI help them with source reduction? If the data submission process were simpler and less time consuming, would facilities be more likely to view the process positively? We return to these important questions in chapter 7.

Facility-Group Interactions and Environmental Performance

Green facilities communicate less with suppliers about pollution control than do other facilities. At first this may seem contradictory (why wouldn't green facilities communicate more, not less?), but it makes sense given the nature of the data. Facility performance was evaluated between 1991 and 2000. Our survey and most of the interviews took place in 2005. We would argue that green facilities have a track record that makes it less necessary to talk with suppliers on a regular basis. The green facilities and their suppliers would likely already have a clear working relationship and expectations about minimizing toxic pollution when possible (Fiorino 2006, 115) The motivation to perform when it comes to release and risk reductions may not be less, but it would be systematized in a way that is less true for other facilities.

A similar finding occurs when looking at the communication between facilities and trade associations. Non-green facilities communicate more with trade associations about matters of pollution control. Green facilities would not need to talk as much because of their advanced performance levels (Fiorino 2006, 116).

In much the same way that occurs with suppliers and trade associations, green facilities also talk less with their customers or end users about pollution control. Green facilities would not have to do as much hard work to convince customers that they are acting in an environmentally responsible manner.

The results above are all suggestive of a likely time dynamic. Different environmental management behaviors from facilities could be expected at different junctures, which in turn would be likely to influence environmental performance differently. Green facilities might have had more intense interactions related to environmental performance at an earlier stage, while brown facilities would be more likely to communicate around environmental performance in more recent years. This may dovetail as well with increasing or decreasing emissions. In particular, as noted above, brown facilities saw their emissions increasing over the 1990s. These increases may have been the motivation for increased communications to reverse course in the 2000s.

As mentioned above, all facilities tend to have limited contact with local emergency planning committees (LEPCs), environmental organizations, regulators, legislators, and the media. Green and non-green differ in no statistically significant way in these aspects, in part because the overall level of such communication is so low.

When it comes to all types of interaction, whether with customers or legislators, brown facilities differ in no statistically significant way from other facilities. Looking at green and brown facilities exclusively, there is some slight evidence to suggest that brown facilities interact more readily with customers, trade associations, and corporate management than do green facilities. The evidence is not statistically strong ($p<0.10$), but it is in line with some of the results mentioned above.

Being a green facility has some limited benefits when it comes to media interactions. Green facilities, in comparison to their counterparts, have seen mass media coverage of their TRI releases decrease to a greater extent than other facilities. It is important to point out that a fair number of facilities ($N = 108$), both green and non-green, get no media coverage whatsoever. But of those that do ($N = 129$), being green translates into getting less coverage over time. Most businesses are likely to see decreasing coverage as a positive, because businesses in general would fear negative coverage about toxic releases. As Hamilton (1995) observed, journalists are more likely to write about facilities with higher levels of emissions. When those emissions go down, we would expect news coverage to go down as well. In contrast to changes over time, green facilities and other facilities do not seem to differ significantly when it comes to their evaluations of the extent of current media coverage or the favorability of that coverage.

Brown facilities seem not to differ in media coverage from all other facilities or even from green facilities in particular. In some respects this is surprising because generally we would expect poorer performers to gather more media attention. The potential explanation could relate to the point made above about how large emitters these facilities are in the first place. Since brown facilities were generally smaller emitters in the early 1990s, it may be that as a group they were simply below the media's radar screen. Our case studies reinforce this conclusion; a number of small facilities made clear that they work hard to avoid calling attention to themselves. This is because they view any visibility of their lagging environmental performance as coming with significant risks of regulatory oversight and possible enforcement actions.

TRI Impacts on Facility Environmental Management
Green facilities and brown facilities varied in a few ways when it came to the TRI impacts on facility environmental management.[5] First, and importantly, it was the brown facilities that found some practical uses for the TRI. Brown facilities were more likely to find that their experi-

ence with the TRI allowed them to gather more accurate data or better estimate releases and allowed them to check their release values against applicable permit limits. This result makes sense. Green facilities have less to learn from the TRI, whereas brown facilities would be more likely to gain some knowledge from their own TRI reporting. Brown facilities need to figure out what is going wrong with the trends of their emissions and begin to change these patterns. Some of our survey respondents back up this point. Green facilities made the changes they needed to make a decade ago, and are now not so receptive to the TRI. As one respondent put it, "We have found that the public is much more interested in companies that are well managed and focused on reducing their impact on the environment. We find that examples of good environmental performance and open honest discussions are much better ways to evaluate company value than TRI reporting." Or as another facility respondent pointed out: "Significant reductions were made early on when the TRI program was implemented. [Our company] proactively seeks alternatives and implements nontoxic alternatives and chemical processes and source reduction measures during the process/product development stage." There was also weaker evidence ($p<0.10$) to suggest that brown facilities were more likely to find that their experience with the TRI gave them a better understanding of the costs and benefits of dealing with chemicals. Brown facilities also were more likely to see their TRI experience helping to increase their capacity for emergency management.

The survey results clearly suggest that the TRI does seem to affect some facilities, but that impact does not translate into clear patterns of performance most of the time. Some green, brown, and mixed facilities found that the TRI affected their ability to demonstrate a commitment to release reduction, and some did not. This lack of a pattern across different types of facilities is as important as any patterns we do find. It suggests that many of the variables we are testing are not very strong in shaping environmental performance, and therefore cannot account for much of the variation among facilities.

Facilities positively affected by the TRI program (regardless of whether they were green, brown, or mixed) were also facilities that had more extensive media coverage and a better overall experience with the TRI; in addition, they were facilities that had greater interactions with suppliers, trade associations, and corporate management.[6] These facilities were ones as well that were strongly driven in their toxic chemical releases management by factors such as a desire to improve their regulatory compliance and having an interest in strengthening their firm's reputation. Whatever their performance, the experience of the facilities with the TRI program connected strongly with their mindset about the management of toxic chemical releases.

Toxics Release Management Factors and Environmental Performance

When asked what factors are important to the facility's management of toxic chemical releases, green facilities were less likely to answer with a desire to improve environmental performance (see table 6.1 and figure 5.3 for detailed results). This makes sense: why prioritize improving environmental performance when you are already a leader in doing so? Other priorities can come to the forefront once strong environmental performance is the baseline.

This result for environmental performance is striking. It suggests that once certain performance criteria are being met, facilities might prioritize other concerns in turn. What is also striking is that green facilities and non-green facilities do not differ in how they rate the importance of other factors in influencing their management of toxic releases. Whether rating the importance of minimizing legal liability or their desire to improve relations with the local community, greens and non-greens were quite similar.

Consistently brown facilities are more driven than other facilities to improve employee motivation in environmental management, minimize legal liability, and strengthen the overall reputation of their firms. These results suggest that facilities that are consistently brown realize the weaknesses of their environmental performance and actively hope to avoid being seen as significant polluters.

Conclusions

The results above can be understood as partial answers to the research questions set out at the beginning of this chapter. The implications of the data are mixed. First, there is limited evidence to suggest that as facility capacity increases, facility performance improves. Facility expertise does not seem to distinguish green facilities from other facilities. The length of time in position for respondents, the number of overall employees, the number of EHS employees, and the level of professionalization all varied similarly between green facilities and between other facilities.

When it comes to management commitments to environmental practices, green facilities and other facilities differ in some ways, but not in others. Green facilities are less likely than others to have specific objectives for air emissions. Although this can be perceived as a symbol of less commitment by green facilities, an alternative interpretation could be that green facilities, by the very nature of their past performance, need not focus on air emissions so closely given their track record. At the same time, the fact that green facilities and other facilities are equally likely to have an environmental management system and/or ISO certification in place, suggests that management commitments to environmental practices, or lack thereof, hold steady despite variations in facility performance.

Green facilities and other facilities both express a wide variety of views when asked about their level of satisfaction with the TRI program. A larger number of respondents were either positive or had mixed feelings toward the program. Only a small percentage felt consistently negative about the experience.

Second, there is limited evidence to suggest that citizen or community capacity affects facility performance. To the extent that local civic capacity can be measured by the level of interaction between facilities and community groups, not only are such interactions infrequent for all facilities, but they are no more frequent for green facilities than other facilities. There seems to be a disconnection between community action and facility performance. This may be explained in part by the time period we are analyzing through our surveys. The community interactions may have occurred in the early years, setting the course for later years. As some of our illustrative cases reminded us, even one occurrence of negative press coverage can make an impression on a facility. In one case, a Midwestern company got hit once with an environmental fine and received negative and most unwelcome press coverage. Since that point the facility has been determined to not let it happen again. This is yet another reminder that the effects that information disclosure programs like the TRI have on facilities is closely tied to related experience under state and federal regulatory programs. The facilities very much see them as related. Experience with one affects the way the other works.

Although neither green facilities nor other facilities had extensive media coverage, when looking at trends over time, the media attention to green facilities slightly decreased, while for other facilities it slightly increased. Although greater initial attention from the media did not translate in better facility performance, those facilities that did decrease their releases and risk levels also saw decreases in media attention. This latter result makes sense and could be seen as a corollary to our original hypothesis about media attention.

In sum, only a limited set of factors appear to correlate with environmental performance. Yet even these factors, combined with what we have learned about facilities in general and about variations across states, can inform the way we think about the TRI program, about information disclosure more broadly, and about where we go from here.

7
Conclusions and Policy Implications

The 2009 TRI Public Data Release reported the release of 4.1 billion pounds of toxic chemicals in the United States. This represented a drop of 5 percent from the previous year and reflected a continuing trend of decreasing releases as a result of pollution prevention, changes in industrial use of chemicals, and a decline in U.S. manufacturing, among other factors (U.S. EPA 2009). But the 2009 data release was distinctive in at least one respect. It was the first to be prepared under relaxed requirements set in place by the George W. Bush administration and finalized by the EPA in January 2007 as the TRI Burden Reduction Rule. The new rule allowed smaller producers of toxic chemicals to report their releases on a simplified form that provided fewer details than previously, although in 2009 only about 2 percent of the 22,000 companies reporting made use of that option. The agency continued to ponder yet another proposed change, dropping the annual reporting requirement in favor of biennial submission of data. In this case, the EPA said that companies might save $1 million annually in reporting costs. Yet a spokesperson for the American Chemistry Council said the savings to industry would be much higher because companies would not have to estimate or calculate their releases in addition to not having to complete the form. He put the cost of current compliance with the TRI program at $650 million annually, and hence he anticipated a much greater savings than the EPA estimated.[1]

The Bush administration policy shift was short lived. A dozen states sued the EPA over the simplified form rule change, and the U.S. Government Accountability Office (GAO) expressed concern that poor and minority communities would be especially affected by the new rule, in effect saying their residents would be denied valuable information that could assist in meeting EPA goals for environmental justice.[2] In addition, out of concern for potentially lost information that could affect the conduct of epidemiological studies, the EPA's Scientific Advisory Board (SAB) sent a letter in July 2006 opposing the change.[3] For its part, the EPA denied that the two proposed rule changes would cause the loss of much information. For example, in January 2006, Kimberly T. Nelson, then the EPA's assistant administrator for the Office of Environmental Information, said of the proposed switch to the short form that the agency "would get 99 percent of the information we get today. There is less detailed information, but it's like a rounding error on all the emissions we currently collect." She went on to make clear that the EPA was not "gutting" the program as some of the critics had suggested.[4]

Congress considered legislative responses to the policy shift, particularly how to ensure that TRI reports assist in achieving environmental justice as well as the broader environmental reporting goals of the program. In the end, it overturned the Bush rules via a provision that Senator Frank R. Lautenberg, D-N.J., inserted into a budget bill in early 2009; Lautenberg was one of the authors of the 1986 Superfund Amendments and Reauthorization Act that established the TRI program. Lisa P. Jackson, the Obama administration's newly appointed EPA administrator, applauded the congressional move, saying that Lautenberg's provision would restore "the rigorous reporting standards of this vital program" (Wald 2009a). As a result, the TRI program was slated to continue as it has operated in prior years even if additional reform proposals can be expected.

Debate over the value of the TRI program and the information that it provides is part of an ongoing and broad reassessment of U.S. environmental policies that shows little sign of dissipating even if only limited political consensus exists on what ought to be changed and how (Eisner 2007; Fiorino 2006). After four decades of reliance on command-and-control regulation, information disclosure and other new policy approaches increasingly are receiving serious consideration by policy scholars and policymakers. Yet empirical analyses of their effectiveness, efficiency, and equity remain limited (e.g., Bennear 2007; Coglianese and Nash 2001, 2006a, 2006b; Dietz and Stern 2003; Woods, Konisky, and Bowman 2009).

As we argued in chapter 1, the next decade inevitably will bring proposals of many new ideas, at least some of which will lead to policy changes. These are likely to include reforms of the existing core environmental protection statutes, such as the Clean Air Act and Clean Water Act, and a variety of measures to deal in innovative ways with energy use and climate change, transportation, urban design and building standards, land use, agriculture, and sustainable development (Kettl 2002; Schoenbrod, Steward, and Wyman 2009).

The EPA's announcement in 2009 of its new mandatory greenhouse gas (GHG) reporting requirement is a case in point. With the United States continuing to debate the specifics of a national climate change policy, the agency's action was a telling sign that information disclosure was alive and well as a policy strategy. The agency noted the value of such an inventory to the business community, which is likely to face both regulatory and economic pressures to reduce their GHG releases, and to the American public. Collection of the data, the EPA said, will "allow businesses to track their own emissions, compare them to similar facilities, and provide assistance in identifying cost effective ways to reduce emissions in the future," even as the public gains "critically important knowledge" that will help in determining "how best to reduce those emissions." Such rationales are similar to the expected utility of the TRI program, and which our survey data tell us is indeed helpful both to industry and to federal and state regulators.[5] In light of these similarities, some of our recommendations in this chapter for how to improve the TRI program may also apply to the new greenhouse gas reporting requirement. As businesses also begin reporting on sustainability initiatives, some may choose to integrate these various measures into a consolidated sustainability report. In addition to such parallel applications within the United States, we hope that our findings about the TRI program speak as well to how comparable information disclosure efforts in the European Union, Canada, and other settings might be designed, implemented, studied, or reformed.

Some of the most creative and promising environmental policy actions, relying on information disclosure and many other strategies, already are evident at state, local, and regional levels, and the national political climate may well create significant opportunities for advancing environmental policy in the years ahead (Klyza and Sousa 2008; Mazmanian and Kraft 2009; Vig and Kraft 2010).The same might be said for similar programs in other countries or international efforts to promote information disclosure.[6] As a result, there is certain to be a lively conversation over the merits of these ideas, and advances in policy research that speak to the impacts of present policies and the promise of proposed reforms. The TRI program is particularly suitable for such assessment in light of its reliance on information disclosure as a policy strategy and the praise it has received for its effectiveness in reducing public exposure to toxic chemicals. The program also has been in existence long enough to permit at least some systematic analysis of its achievements and effects, both on corporate environmental performance and on residents of communities in which industrial facilities are located.

Scholars in public policy, law, political science, sociology, risk communication, and management have contributed much to our understanding of the TRI program and information disclosure as a general practice. This research has informed our understanding of the origins, purposes, and overall impacts that the TRI program has had since its inception in the late 1980s (Fung, Graham, and Weil 2007; Graham 2002; Hamilton 2005; Herb, Helms, and Jensen 2003; Hadden 1989; Weil et al. 2006). Moreover, given the enormous TRI database, scholarly analyses have been able to offer insights into industry compliance and reporting over time, and the factors that seem to drive such compliance (Grant and Jones 2004; Grant, Jones, and Trautner 2004; Santos, Covello, and McCallum 1996), overall trends in and impacts of the release of toxic chemicals (Hamilton 2005), the implications of these releases for environmental justice (Abel 2008; Ash and Fetter 2004; Pastor, Sadd, and Morello-Frosch 2004; Shapiro 2005), and some confirmation of how much industry has improved its environmental performance as measured by changes in chemical releases (Coglianese and Nash 2006a; Press 2007).

We know much less about citizen or community knowledge and use of the TRI reports, where only a few limited surveys have been conducted (Atlas 2007; Beierle 2003; Lynn and Kartez 1994; U.S. GAO 1991), and not much more about how the TRI program has affected chemical management within industry. In a telling reminder of these knowledge gaps, a 1993 EPA report on how the TRI data have been used consists chiefly of anecdotal evidence of use in selected communities and states and a brief summary of the few scholarly studies available at that time (U.S. EPA 2003). To understand better how the TRI program has fared, the mechanisms through which it has worked, and the effects it has had both within industry and communities requires a somewhat different kind of empirical framework than has been employed previously. We hope we have contributed to advancing this understanding through both our performance dilemma typology and the multiplicity of methods employed in this analysis, even if we have opened the door only a little wider in trying to answer some of the major questions that should be asked about information disclosure programs and their effects.

We think that the data we have collected permit us to speak with at least some confidence about what the TRI program has achieved, where it has fallen short, and some of the factors that make a difference in moving facilities and governments beyond the performance dilemma that we described in chapter 2. Similarly, the analysis suggests what might be done to increase the effectiveness and efficiency of the program, and to provide the public with the kind of information that can allow informed judgments about acceptable health risks, spur manufacturing facilities to further improve environmental performance, and complement the array of other environmental policies. In the rest of the chapter we review our major findings, point to future research needs, and examine the policy implications.

Overall TRI Trends

Since the inception of the TRI program, core chemical releases have decreased substantially, and they have continued to decline in the most recent reporting periods. Our particular subset of 8,389 facilities reporting in 1991, 1995, and 2000 mirrors the overall pattern for the 22,000 plus facilities reporting each year. As we discussed in chapter 1, looking only at core chemicals, total on- and off-site disposal or releases for all facilities declined by 61 percent, or 1.83 billion pounds, from 1988 to 2007. For our sample of facilities, toxic air releases went from a combined 1.21 billion pounds in 1991 to 965 million pounds in 1995 and 677 million pounds in 2000; this represents a reduction of 44 percent for the ten-year period.

In light of trends like these, the TRI program could be declared a success. Facilities have responded to the requirements for information collection and dissemination by making available vast quantities of data about chemical releases over more than two decades and also by making significant changes in their management of toxic chemicals to reduce both the quantity of releases and the public health risks they impose on surrounding communities. These outcomes suggest that information disclosure as a policy strategy can work. In this case it has produced unique and valuable information that has been made widely available and has been used in a variety of ways to improve corporate environmental performance (U.S. EPA 2003)

These findings reinforce the prevailing view that the EPA itself has long held about the program and its achievements. The agency said in its 2007 TRI report, for example, that the program "provides the public with unprecedented access to information about toxic chemical releases and other waste management activities on a local, state, regional, and national level." It went on to argue that the program helps to "identify potential concerns and gain a better understanding of potential risks," "identify priorities and opportunities to work with industry and government to reduce toxic chemical disposal or other releases and potential risks associated with them," and "establish reduction targets and measure progress toward reduction goals" (U.S. EPA 2007).

Yet these achievements and benefits of the TRI program are by no means uniform. They vary considerably across industrial sectors, states, communities, and individual facilities. As we showed in the previous chapters, there are leaders and laggards in the management of toxic chemicals, with many facilities and firms showing remarkable improvements in environmental performance while others exhibit no improvement at all or declining performance. This variation has been insufficiently emphasized in discussions of the TRI program and of information disclosure in general. The EPA and independent analysts have focused on the aggregate trends across all manufacturing industries, a practice that tends to give a misleading picture of how facilities are performing. We believe it is essential to understand the variation among facilities and states, why it exists, and what difference it makes.

A major implication is that any assessment of the TRI program's impacts and any suggestions for policy change designed to improve its effectiveness must take this variation into account. Clearly, one size does not fit all. Larger and smaller facilities, for example, have widely differing capacities for data compilation and reporting and often for pollution reduction activities. Facilities in states with strong regulatory programs will likely respond differently to the present incentive structures than will facilities in states with weaker, less consistent, or poorly enforced regulations. Facilities in communities with vigilant environmental and citizen groups (or strong local media coverage of environmental and public health stories) likely will face differing demands compared to those in other communities. As reported in chapter 4, for example, industrial facilities in states with a stronger presence of environmental groups had a higher level of environmental performance. Similarly, facilities in states with a stronger commitment to environmental protection (as indicated by spending on hazardous waste programs) were more likely to have a better record of environmental performance. And facilities will be more likely to become cleaner and safer in states signaling a cooperative commitment to performance with pollution prevention assistance. In sum, facilities and the states in which they are located are sometimes likely not only to vary in their ability, but also in their motivation to move beyond simple compliance and move toward a performance orientation.

The literature is mixed on these kinds of relationships. For example, Grant and Jones (2004) found, like us, that state-level right-to-know programs had no significant net effect on toxic chemical emissions from facilities and they question the efficacy of information disclosure laws. However, it may be that to study the relationship requires a better measure of state action (such as funding levels or other implementation actions such as inspections, auditing, and enforcement) than simply having or not having a right-to-know program. Also, when many different facilities are aggregated, some critical information can be lost. In part for that reason, we offered assessments based on a performance typology of facilities (green and brown facilities, for example) that can better capture the wide variation among industrial facilities. Past versions of the regulation (Potoski and Prakash 2004) or enforcement dilemma (Scholz 1991, 1984) juxtaposed evasion versus self-policing for facilities, and deterrence versus enforcement for regulators. Such a framework is useful for the analysis of first and second generations of environmental policy outcomes, but falls short in framing many of the experimental policies emerging in today's landscape.

We therefore offer a modified performance version of the classic collective action problem in environmental policy to use in future research. We also followed Daniel Press's (2007) recommendation noted in chapter 4 that analysis ought to make good use of comparative studies that can point to real differences in environmental outcomes as well as nested studies that make use of both large-N quantitative analysis and the insights that small-N studies can provide. While our state analysis looked at 48 cases of the trends in over 8,000 facilities, future case studies with even smaller samples can track how management decisions change over time in response to varying conditions. Small-N studies also would be ideal for improving our understanding of the role of community activists and how information such as that provided by the TRI is communicated and used within communities, and how social networks contribute to the process. Our performance dilemma framework could inform future social science research on environmental performance.

Strengths and Weaknesses of the TRI Program

Almost any public policy program has weaknesses as well as strengths, and the TRI program is no different. Despite the overall downward trend in releases and risks, critics have pointed to the limited nature of TRI data, the cost and burden of collecting and disseminating the information, and the apparently limited use made of the inventory, particularly by the general public (Atlas 2007; de Marchi and Hamilton 2006; Graham and Miller 2001; Natan and Miller 1998). Yet supporters of the program, especially the EPA, see it as an effective way to alert the public to community health risks and to track progress over time in dealing with toxic chemical management even if they do recognize that improvements can be made in the program.

Our surveys of both corporate and public officials convey a similarly mixed picture, but one that is far more positive than negative, with some important implications for policy reforms. As we noted in chapter 5, some 36 percent of the corporate officials we surveyed viewed the TRI program as positive or strongly positive, about 50 percent saw it as mixed or neutral, and only about 12 percent viewed it negatively. To put the results slightly differently, 86 percent of our respondents were *not* negative about a program that critics fault for being a costly burden on the business community, and that the Bush administration targeted for "burden reduction." Curiously, the worst performing facilities (brown facilities) were more satisfied with the TRI program than were the best performing (green facilities), somewhat contrary to what one might suppose (see table 6.1). Yet we also found that neither brown nor green facilities communicated much with local community groups. Regardless of a facility's level of satisfaction with the program or perception of its burdens, much more might be done to improve communication with community residents. In so doing there may be more incentives created for facilities to see regulations as only a baseline from which they could further improve their environmental performance.

When asked about the effects of the facility's experience with the TRI program, the picture was similarly mixed. Most facilities said their experience over time allowed them to better understand what they needed to address in subsequent TRI reports, to set goals for or demonstrate commitment to release reduction, and to better understand the costs and benefits of dealing with toxic chemicals. Not surprisingly, the sample was more divided on whether the program helped them to reduce community complaints or expressions of concern, identify needs and opportunities for source reduction, or increased their capacity for emergency management. Few facilities stated that the TRI reports increased their ability to discuss chemical releases with the local community and media. Indeed, as we discussed in chapter 5, most reported very little media coverage, and our questions on the role of the media received very high "don't know" responses; that is, the facility managers were not aware of much media coverage and thus could not characterize it. These findings are consistent with their reports that they have very little interaction with the media, with legislators, with local emergency planning committees, or with community or environmental groups. However, with a higher level of community interest and greater communication with local facilities, we believe that facility managers would be moved to improve their environmental performance. We make some suggestions later in the chapter for how communication and the public understanding of chemical releases and risks could be improved.

Other survey findings also suggest a more accommodating view of the TRI program and of parallel federal and state regulatory programs than one might imagine from press accounts. For example, the overwhelming majority of public officials at all levels of government thought that the state political climate was supportive of business, although smaller percentages characterized the working relationship between industry and federal pollution control officials as good (nearly 50 percent of federal and state officials, and 43 percent of local officials). The percentages were a little higher when the question emphasized the working relationship with state pollution control officials (nearly 62 percent for federal and state officials and 50 percent for local officials). Nearly 80 percent of local officials said that industry works well with local government on pollution control, in sharp contrast to the views of federal and state officials, only 45 percent of whom agreed. Similarly, local officials believed that industry is honest and trustworthy on matters of pollution control (57 percent) whereas state and federal officials were somewhat less positive (45 percent). At the same time, local officials are much less interested than state and federal officials in the TRI program and the data that are collected. *We think that much more could be done to bring local officials into the process of chemical management.* As we suggest below, one way to do that is to transform the TRI data into a more useful metric that offers a clear understanding of community environmental and health risks, as the RSEI model does to some extent.

As we reported in chapter 5, most federal and state officials also indicated the considerable value of TRI data as a tool with which they worked on pollution problems in their geographic area (62 percent). They used the data in diverse ways, including educating citizens about local pollution problems, assisting with regulation and enforcement, increasing their knowledge of local pollution problems, comparing emissions to similar facilities, comparing facility emissions over time, checking facility emissions against permit records, identifying needs and opportunities for pollution prevention and source reduction, and setting local, state, and regional environmental priorities. However, relatively few reported exerting pressure on business, assisting citizens in negotiations with facilities, or even contacting businesses or the media.

As reported in chapter 4, we found that state-level disclosure programs appear to make little difference in the overall environmental performance of facilities, but pollution prevention programs at the state level did have an impact. One explanation may be that pollution prevention programs involve much more direct contact between facilities and state regulators, or are taken more seriously by the facilities whereas the state disclosure programs do not represent an added incentive to improve performance. As we discussed in chapter 4, pollution prevention may be an important cooperative signal instead of the typical compliance tack from state regulators that modifies the incentive structure of the performance dilemma.

Taken together, these findings indicate a generally positive view of the TRI program on the part of industry as well as federal and state officials, and substantial utility of the program's data as part of regulatory activities even if the information is not used as much by the public itself as might be imagined. In some ways these findings are encouraging. The media may no longer cover TRI data release as much as they used to in the late 1980s and early 1990s. However, the regular updating of facility releases hardly goes unused. Moreover, voluntary programs like pollution prevention require a reliable inventory of pollution volumes and the TRI, according to our survey, has enhanced the knowledge of both facilities and government in that regard.

Use of TRI Data

Based on our survey results, it is clear that the kind of information the TRI provides would be more useful to facility managers, public officials, and citizens if it could be presented in ways that better clarify relative public health risks and are more easily understood, particularly for non-professionals at the community level. For example, an online and easily accessible database that could model and estimate individual risks by location might attract more public and media attention and thus potentially galvanize pollution reduction activities. One can even imagine that at some point a neighborhood's risk factors might become part of the expected information to be made available when individuals search for homes to purchase or apartments to rent, much as is now the case with risks such as location within a floodplain, the existence of lead-based paint, or contaminants in drinking water.

The potential as well as the pitfalls of releasing TRI information in this way was demonstrated in late 2008 when *USA Today* used the EPA's RSEI model to report on cancer and non-cancer health risks in the air surrounding 128,000 schools throughout the nation. Any of these schools could be quickly located by entering the city and state in the newspaper's Web page that drew from the study's database. The report concluded that "the potential problems [of chemical risks] that emerged were widespread, insidious, and largely unaddressed," even as critics questioned the methodology that was used and in many cases on-site testing that followed public outrage revealed no measurable levels of toxic chemicals at the schools.[7] The study was merely one of the most recent reminders that when presented in a certain way (e.g., simple, clear, and easily accessible), the public is eager to have this kind of information and the local press is often willing to cover the story. The downside of such a report is that the information may be incorrect or subject to misinterpretation, leading to unfounded public fears and inappropriate actions. The challenge is to develop a suitable mechanism for such information disclosure that is both accurate in its risk estimates and that can point to ways in which local citizens can work to resolve the problems that are identified, for example through community partnerships with government agencies and corporate facilities.

The present TRI database leaves much to be desired in these respects, and serious consideration should be given to its improvement. There is much that might be done. Providing the data by facility in pounds of releases is not a very meaningful metric to most citizens. Nor would some risk measure, derived from RSEI or an improved model, be sufficient if it were not possible to citizens to understand what the numbers mean for them and their families. It should be said, however, that even a very small percentage of knowledgeable and active citizens in a community (e.g., 1 percent or less) may nonetheless translate into a substantial number of people whose actions can influence corporate management as well as decisions by public officials. Indeed, when the group Environmental Defense Fund launched its Scorecard Web site in April of 1998, based heavily on TRI data *transformed into cancer risk quantities*, it claimed that public response was "overwhelming: hundreds of thousands of citizens clearly wanted information about pollution in their community and were eager to take action to defend their environment." The group's appraisal may well be correct despite the findings of surveys suggesting that only small percentages of the population seek out TRI data (Atlas 2007). The Environmental Defense Fund expanded and refined its Web page over seven years, and argues that it became "the web's preeminent source for information about toxic chemicals." In 2005, the group turned the site over to another nonprofit organization, and its utility seems to have diminished since that time, primarily because it was not being updated regularly.[8]

It is equally important that such information actually include the full range of toxic chemicals that facilities release that can affect public health. The current TRI does not do this. It remains limited because of the number of substances omitted from the list of reportable chemicals and because of the thresholds that apply—exempting many facilities from the need to report at all. Of course, there would be an added cost to lowering the threshold or increasing the number of chemicals that must be reported, and such costs are likely to be strongly resisted by industry. An alternative is to make available to the public information that is now collected under the Clean Air Act's permitting program through National Emissions Inventory, but it would have to be in a form that is easily accessible and understandable to the general public.

Moreover, having RSEI data available only on a CD-ROM has made little sense in recent years. Fortunately, the EPA is now making the data available by download directly from a dedicated Web site.[9] Yet the agency also could make it easy for users to find pertinent information by community or zip code, or via an interactive map of facility locations, releases, and risks. Of course, all of the usual qualifications and limitations that the EPA announces with distribution of the TRI data would have to be prominently displayed and caution urged for the use of the data.

When the TRI first became available in June 1989, it was the most popular of all the online government databases on toxic chemicals—at a time when the Internet was in its infancy. But that alone did not mean that citizens were able to use the information effectively. Early studies of public knowledge of the TRI, its use, and facility communication with the public hinted at limited public interest and only modest use of the data in the late 1980s and early 1990s. For example, a GAO study from 1991, shortly after the TRI reports began, found that most TRI facilities (83 percent) reported only "mild or weak" public responses to the availability of TRI data. The GAO also found that some 77 percent of the

TRI facilities that were surveyed had received no direct inquiries for TRI information. In their 1990 survey of a small sample of industrial facilities, Lynn and Kartez (1994) found that 79 percent rarely and 21 percent occasionally received requests for TRI data.

Somewhat more encouraging, as part of its 1991 study, the GAO conducted a limited survey of 500 adults in three counties with substantial TRI emissions and reported that 47 to 58 percent of respondents in those counties "had read or heard reports" about local emissions and 15 to 20 percent said they knew that toxic chemical emission reports were publicly available (U.S. GAO 1991). The GAO study also found that most of the use of TRI data in 1991 was by industry and research scientists rather than by citizens or state agencies, but this pattern appears to have changed as online access improved over time.

In a 2001 study of two counties in North Carolina and Maryland, Atlas (2007) found that only a relatively small percentage of people were aware of the TRI program and even fewer did anything as a result. Indeed, he found no improvement in public knowledge of the program compared to the 1991 GAO survey despite twelve years of experience in making TRI data available via an increasingly accessible Internet. Limited public knowledge of a technical program like the TRI in many ways is not a surprise. Public knowledge of environmental problems and policy issues has never been very high, and the issues persistently rank among the least salient of contemporary concerns (Guber 2003; Guber and Bosso 2010). Nonetheless, if reflective of experience in other parts of the country, these findings hint at some potential barriers to information disclosure efforts of this kind.

As we suggested in our review of the corporate survey results, facilities often anticipate a public or media reaction to release of less than positive information, and hence attempt to improve their environmental performance even in the absence of overt citizen action. It is the anticipation of negative press coverage and public outrage that makes the difference. Still, that mechanism depends on the release of regularly updated information to the public, and a sufficient number of interested and active citizens to make the threat credible. In 2009, the EPA announced that the TRI data would now be made available within one year of their collection from facilities rather than the two-year lag that has prevailed for years. By December 2009 the agency did indeed keep to that schedule as it released the 2008 TRI data in the same year that it received the information from facilities.[10] This is a distinct improvement and it will surely help to ensure that citizens, administrators, and other interested parties have somewhat more current data. *Yet we think the purposes of the TRI program would be best served by an even shorter turnaround time, if possible, which will help to ensure public access to the most current data.* We believe it will become increasingly feasible to do so as TRI data are submitted to the agency electronically.

The EPA should also release TRI data in a manner that permits an analysis of facility environmental performance changes over time as we have done here. The current TRI provides essentially the same kind of information that has been released since its launch over twenty years ago. While trend data on toxic releases can be pulled from the TRI, the agency might consider altering the manner in which the information is displayed to make it easier for readers to detect positive or negative changes in environmental performance in either releases or risk levels or both. As we discussed in chapter 1, other kinds of information disclosure programs suggest the possibilities.

A related question is whether the TRI program has outlived its usefulness, as Atlas (2007) suggests. In the late 1980s, there was little regulation of air emissions of individual chemicals, and thus the TRI filled an important gap. After the Clean Air Act Amendments of 1990, however, the federal government and the states formally regulated hazardous air pollutants via a permitting system and made available data on emissions of hazardous or toxic air pollutions. While regulation via environmental protection statutes potentially could reduce the need for TRI generated information, the reality is that the EPA has been slow to implement the air toxics program. According to the GAO (2006), the EPA has made only modest progress, most of its regulatory actions have been late, and major components of the program have "still not been addressed." The EPA's emphasis has been on large stationary sources rather than on smaller and mobile sources. In addition, the agency has suffered from litigation over the program, often misses deadlines, and, according to the GAO, it "lacks a comprehensive strategy for completing the unmet requirements or estimates of resources necessary to do so." It would seem that the program has been a low priority for the EPA. However, as noted just above, the EPA could consider making available to the public data collected for this program in the National Emissions Inventory, which should require a relatively modest investment.[11]

The TRI program could be made more relevant and useful to facilities if it could be linked to their use of environmental management systems (EMSs), such as the ISO 14001 set of standards. We reported in chapter 6, for example, that brown facilities are more likely to have an EMS than all other facilities. If such EMSs specifically addressed collection and dissemination of TRI and related data, they might assist facilities in more accurately accounting for their pollution releases and in finding ways to reduce releases over time. Stronger connections between TRI data and EMSs would also serve to strengthen the role that the TRI can play in creating incentives for better environmental performance. Like the EPA's effort with the pollution reduction challenge 33/50 Program and Project XL (for excellence and leadership), the agency could pilot a performance version of the TRI with receptive facilities. This is not to say that the TRI would no longer be used to supplement traditional command-and-control actions, as our survey indicates is the case. But for the TRI to maintain its relevance it must serve the interest of facilities as well as those of public officials, and linking the inventory to facility EMSs would help to do that.

What might be done to increase public interest in the issues and to facilitate public access to potentially useful information? In her 1989 book, *A Citizen's Right to Know*, Susan Hadden suggested many ways to improve citizen understanding and use of the then new TRI data, including use of community workshops on toxic chemicals, citizen advisory committees, workplace education programs, and the mass media. Today there may be similar potential in the use of social networking sites such as Twitter, Facebook, and MySpace, or blogs and wikis. Individuals would be less dependent on their own capacity to interpret the information and could benefit from exchanges with others who have greater experience, knowledge, and time to review such data releases.[12] *We think the EPA should give serious consideration to the creative ways in which citizen understanding and use of TRI data might be improved.*

It is difficult to determine the extent to which such suggestions were acted on by either industry or government, but the EPA has experimented with community workshops and for some time has funded the development of partnerships between communities and local industry.[13] In addition, the EPA's Web site for the TRI program indicates that substantial efforts have been made to offer training workshops for facility staff involved in reporting TRI data, to initiate stakeholder dialogs that focus on reduction in the burdens and costs of TRI reporting (for example through revision of reporting forms and software), and similar actions. Web sites of regional EPA offices indicate a comparable level of concern and activity.[14] For its part, industry has made similar efforts. The American chemical industry has sponsored hundreds of citizen advisory panels across the nation to try to rebuild public trust in local facilities (Lynn, et al. 2000). Still, much more could be done to reach out to the public and to create a more informed community that can be involved in decisions about management of health risks, whether through the agency's Community Action to Restore the Environment (CARE) program or a different agency initiative.

Some other changes that might be made to improve the utility of the TRI database are mundane but nonetheless important. As we noted in chapter 5, the TRI dataset is filled with errors and obsolete information. For example, the TRI listing of a facility includes its physical location and contact information, but we found that the location given was not always a proper mailing address. Citizens attempting to contact a company would find that their letters would be returned by the postal service as undeliverable. Calls made to a facility often go answered, and messages must be left on voice mail. But the calls simply may not be returned. Moreover, the pubic contact official at the facility may or may not be listed correctly. We found the number of errors or lack of currency in contact information to be substantial, possibly discouraging citizens who might try to contact a facility over concern about its emissions.[15] For a program that is designed to facilitate public acquisition of information and to promote facility transparency and public accountability, these are significant weaknesses that should be addressed. Moreover, the TRI in its current form is poorly designed for the social learning systems advocated by many regulation critics (Fiorino 2001; National Academy of Public Administration 2000). *Thus we believe that the EPA should find practical ways to improve both the accuracy and currency of the information incorporated in the TRI database, and particularly for information that affects public efforts to communicate with facility management.*

Of course, public databases are used by a diversity of people with many different interests. Even if the general public is not very likely to tap into the database, many others are, including state and federal regulators. The implication is that to the extent that the TRI is altered in the future, it is important to bear in mind that the utility of public disclosure of information cannot be measured solely by direct use by individual citizens. The EPA itself reported that for a typical month in November 2005, the agency's TRI Web site received 7,666 visits. The agency said that the users included medical researchers, investment analysts, the insurance industry, regulators, consultants, states and localities, and the public (Skrzycki 2006). These findings are similar to those of the GAO survey of 1990, where most users had a corporate or scientific affiliation

(GAO 1991). In our own survey, most of the federal and state officials used the EPA's TRI Explorer or Envirofacts, the Public Data Release, or state agency sources, or they contacted industrial facilities directly. However, only very small percentages of local officials surveyed (5 to 15 percent) used any of the official sources.

The U.S. EPA is aware of many of these long-standing obstacles to a fuller use of TRI data. Indeed, in 2009, the agency developed a new Web page in cooperation with the Environmental Council of the States (ECOS) as a collaborative forum for users of TRI and similar data to "vet their analyses, share success stories and best practices, and collaborate on solving community chemical-related problems." The hope is to illustrate how TRI data "enable[s] you to make informed environmental decisions within your community."[16] For years the agency also has sponsored an annual TRI National Training Conference in the Washington, D.C. area in which various stakeholders report on experience with the program and seek to improve it. Yet more could be done to reach out to potential users of TRI data and facilitate their use of the information. To address these needs, the agency might look to a fairly well developed literature on public participation in environmental decisions for what could work in this case (e.g., Beierle and Cayford 2002; Dietz and Stern 2008).

One model for greater citizen involvement is the Superfund program, where for years the EPA has made extensive use of community advisory groups to mobilize citizens to play a role in cleanup decisions. Daley (2007) has carefully studied the role of such citizen groups and technical assistant grants that can support them, and she found that the groups can be very effective in altering the way Superfund cleanup actions take place. She found that the EPA is more likely to choose higher standards in cleanup actions that better protect public health when communities have active citizen groups and technical assistance funding. By extension, one might suppose that the more active local groups are in a community, the greater effect they would have on local leaders, government, and industry. While we were unable to study the role of community activists as much as we had planned, this could be a fertile area for future investigations of the factors that affect corporate decisions on chemical management and similar environmental decisions. Grant, Jones, and Trautner (2004) report that while facilities with corporate management located in another community performed no worse than other plants, these kinds of facilities in communities that had more associations emit significantly fewer toxic chemicals than others. So there is reason to believe that community characteristics do somehow make a difference.

Leaders and Laggards: Improving Environmental Performance

One implication of the substantial variability among facilities and firms nationwide is a need to give more consideration to the differential needs of corporate leaders and laggards, as we suggested in several of the sections above. Leaders may be well motivated to comply and to undertake corrective actions to reduce releases and perhaps risk while laggard facilities may not be so well motivated for a number of reasons. Practically speaking, our survey of public officials suggests that state and federal regulators already are well aware of what facilities are doing and many of them are using the annual TRI data reports to further refine their regulatory activities. The practice varies among the states, but we believe the appropriate strategy at both the federal and state level is to *target those facilities and firms that need greater incentives or technical assistance to reduce releases and risks while simultaneously encouraging, recognizing, and rewarding those facilities and firms that are steadily improving their environmental performance.* This was a key element of the EPA National Environmental Performance Track program before the Obama administration ended it in 2009, and experience with that program should help to inform any new efforts to target both leaders and laggards. Much the same could be said about successful state programs that are similar, such as Wisconsin's Green Tier program.[17] This would be a fitting example of what we earlier referred to as applying a policy learning perspective to improving both agency and corporate decisions on chemical management (Coglianese and Nash 2006b; Fiorino 2001, 2006).

Understanding the factors that most strongly affect the management of toxic chemicals by facilities, such as a desire to minimize legal liability, improve regulatory compliance, improve environmental performance, and save money (the four most important, in our survey, as reported in figure 5.3), should assist regulators in designing such strategies. We have tried to identify some of the best-performing facilities as measured by reduced TRI releases and risk and to ask what led to their success in comparison to others. Both the leaders and laggards have much to tell us about their experience, and their perspectives can help in the design of new policy efforts. We think that future research could go well beyond what we have been able to learn about facility and firm behavior, and we are encouraged by the value of a research strategy that focuses on the best- and worst-performing facilities and firms.

For example, we found that leaders and laggards in our sample differed significantly across a number of important variables. These include communication with corporate management, trade associations, and customers (higher for brown facilities than for greens) and their perception of the utility of the TRI program; brown facilities were more likely to value the TRI program for helping them to check their release values against applicable permit levels and for gaining a better understanding of the costs and benefits of managing their chemicals than were green facilities. This also suggests that the weakest performers could value government assistance like pollution prevention more than typically presumed.

We would like to think that community pressure or a desire to improve a facility's environmental reputation is equally important as minimizing liability, improving compliance, improving environmental performance, and saving money. Yet the survey findings suggest otherwise. This does not, however, diminish the potential for the TRI program (and related chemical information programs) to better inform and mobilize communities and thereby to spur companies to improve their performance. Facility managers clearly did speak of the importance of reputation and a desire to improve community relations. Still, the conclusion we draw is that whatever effects the TRI has on facilities is mostly indirect, and closely associated with the facility's ongoing relationship with federal and state regulators. One implication is that information disclosure policies like this can be successful only when coupled with ongoing and predictable regulatory efforts that companies and facilities take seriously. Without the regulatory backdrop, information disclosure would likely achieve far less impressive results. Our analysis in chapter 4 also suggests that what matters most may be the general political, social, and economic climate in a state rather than the attention that a specific facility gets from the local community or media; this relationship is suggested by our findings on the influence within states of environmental group membership and public liberalism. In light of these findings, we would encourage both the EPA and state regulators to recognize the close linkages between information disclosure and regulatory decision making. In particular, we think that *consistent and predictable enforcement of current regulatory statutes could send a useful message to facility managers who may be trying to anticipate what level of environmental performance will be expected in coming years.*

Another of our findings is pertinent here. Contrary to the popular assumptions about regulation and its burdens, our sample of facility managers indicated a great deal of satisfaction with the current regulatory environment. For example, most believed that state regulatory requirements were highly stringent (71 percent somewhat or strongly agreed) and yet they thought that state regulators were honest and trustworthy (68 percent, with few in disagreement). A majority also believed their facility or company had a good working relationship with federal regulators (69 percent somewhat or strongly agreed and very few disagreed). Similarly, most indicated they had a good working relationship with state regulators (85 percent, again with very few in disagreement) and with local government officials (85 percent). Yet some 40 percent of the sample did not report much in the way of offers of technical assistance by state or federal regulators; these findings suggest that they would be receptive to such assistance if offered. Technical assistance is likely to be especially important for smaller facilities that lack the capacity for chemical management that larger firms often possess. And as chapter 4 showed, pollution prevention assistance programs were one of the most influential factors in leading states for green facilities. *So we recommend that federal and state regulators help ensure that such technical assistance is available to all facilities in need of it and adequately supported with agency resources.*

While we found limited evidence that environmental performance was related to firm size and no evidence that environmental performance was related to financial performance (e.g., profitability), the relationships merit closer study than we were able to give them. It may be that facility-to-facility variation is so great that no one factor, such as corporate leadership, industrial sector, nature of the community or state, or experience with the TRI, can account for very much of the differences. Thus no single policy or administrative strategy is likely to be of determinative importance. However, state officials and regional EPA administrators may have the particularized information they need to take appropriate action on a firm-by-firm basis. They are probably more likely than the federal government to be alert to different local and regional capacities and better able to deal with unique situations. State and local officials are in the best position to use a modernized TRI database effectively. *Therefore we recommend that the EPA, in conjunction with the states, develop a second generation RSEI model that can better assist local officials, such as LEPC members, and citizens in understanding the environmental and health risks of toxic chemicals and in developing suitable responses to those risks.*

The Logic of Hybrid Policies

These findings of regular and extensive use of TRI data by federal and state regulators as well as by the facilities themselves are important. If we looked only at how knowledgeable the general public is about the TRI program or the data, or how much people act on the information disclosure in their communities, we would miss the broader picture. Despite these conditions, the release of TRI data plays a significant role in chemical management in part because of the regulatory system in which the facilities operate. Companies engage in pollution prevention and make many other improvements in environmental management not because they fear the public's wrath upon release of data, but because they know that such improvements either are required by regulations or likely will be in the future, that they may face legal liabilities as a result of their operations, and because they value the company's reputation, both in the community and beyond. Earlier we referred to the practice as anticipatory performance and preemptive self-regulation. In short, the public's influence is important, but it is largely indirect. In this sense, information disclosure programs like the TRI play a very useful role in the overall scheme of environmental protection policy. They do not replace regulation but rather supplement it.

One way to think of the future of environmental regulation, then, is the value of hybrid policies, using a combination of complementary policy strategies rather than relying exclusively on regulation, or replacing regulation with one or more of the alternatives. In many, if not most, cases, so-called voluntary or self-regulation, widely touted in recent years, is unlikely to achieve much on its own, particularly if there is no third-party auditing of management systems and results or no sanctions for falling short of promised performance (Borck and Coglianese 2009; Prakash and Potoski 2006; Potoski and Prakash 2005; Potoski and Prakash 2009). The same might be said for voluntary as opposed to mandatory information disclosure programs. As we noted in chapter 1, mandatory programs such as the TRI differ substantially from voluntary release of environmental information. Voluntary programs, of course, may vary in their design, the level of commitment made by affected businesses, and their effects on citizens and regulators. Some of what we have learned about the TRI program may be transferrable to such nonmandatory programs. But it is also important to recognize the differences between mandatory and voluntary information disclosure.

Whether information disclosure is voluntary or mandatory, there are significant barriers to improving environmental performance through this kind of action alone. Even corporations that have a strong sense of social responsibility invariably are constrained by their fiduciary obligations to shareholders, and thus face real limits on how far they can go if environmental improvements are not mandated by law (Eisner 2007; Vogel 2005). Still, we believe that voluntary programs in general can work well in combination with a clear, predictable, consistent, and effectively implemented regulatory structure; under these conditions, corporate voluntarism has an important role to play (Coglianese and Nash 2006b; deLeon and Rivera 2007; Gunningham, Kagan, and Thornton 2003; Lyon and Maxwell 2004; Morgenstern and Pizer 2007; Prakash and Potoski 2006; Press 2007).

Information disclosure policies, particularly those that, like the TRI, are mandatory, appear to operate in much the same way. By themselves they are not likely to bring about a wholesale shift in product or process redesign or lead to a new commitment to corporate social responsibility. Yet when they are linked to other federal and state regulatory programs, they can contribute significantly to improved environmental management and eventually reduction in public health risks. Relying on information disclosure in lieu of regulation is not as likely to be effective in bringing about meaningful improvement in environmental performance.

The TRI's role as both a continuing instigator of environmental performance improvements and a measure of trends over time cannot be overstated. Broadly, the TRI can serve as an additional factor that has the potential to motivate both public officials and facilities to think outside of the compliance box. Combining what we have said about the performance dilemma with our typologies of green facilities and brown facilities, we see a way to map the two ideas together (see table 7.1). Although there will always be variations in the performance of facilities when it comes to toxic chemical emissions, facilities can be nudged toward better performance. Having fewer brown facilities and fewer mixed facilities can be reasonable goals.

Table 7.1
Performance Dilemma and Facility Performance

	Facility	
	Compliance (Increasing Risk)	Beyond Compliance (Decreasing Risk)
Government Encouraging (Decreasing Releases)	*(Mild Booster)* (Yellow Performance)	*(Performance Synergy)* (Green Performance)
Commanding (Increasing Releases)	*(Minimal Expectations)* (Brown Performance)	*(Corporate Citizens)* (Blue Performance)

We merge table 2.1 and table 3.1 in order to reinforce the connection between our theoretical understanding of the performance dilemma and our empirical results for facility performance. Although not every green facility will improve its chemical management because of the encouragement of government and the facility's own desire to move beyond compliance, in general we would expect green performance to include reductions in both the release of toxic chemicals and their risks. In much the same way, although not every under-performing facility will focus narrowly on compliance and therefore experience the heavy hand of government regulation, in general we would expect brown facilities to increase both their chemical releases and the risk associated with them. Although more research needs to be done on the mixed cases, it is reasonable to believe that when releases and risks are not moving in tandem, there may also be a parallel disconnect between governments and facilities. *Based on our state-level results and the facility surveys, we see clear evidence to suggest that getting beyond the performance dilemma is feasible and can be fostered by cooperative signals from federal and state regulators.*

There is another side to information disclosure that merits attention: the ethical basis of such a policy strategy and possible conflicts between the public's right to know and other social goals, such as national security or a corporation's desire to protect its rights. Few would disagree with the proposition that the public has a right to know about potential dangers in the community. The TRI program grew out of concern that Bhopal-like accidents might occur in the United States and public recognition that industrial facilities would probably not voluntarily disclose many possible dangers. The same concerns exist today.

Consider the link between chemical and other industrial facilities and the risk of terrorist attacks. In 2005, one plant outside of New Orleans had 600,000 pounds of hydrofluoric acid on the premises, exposing nearly the entire population of the city in the event of a major accident or attack. According to the *New York Times*, the plant was hardly unique. It is one of 15,000 nationwide that pose such a hazard. More than 100 of these facilities could expose a million or more people to such a risk. The Department of Homeland Security has alerted the public and policymakers to many other risks of this kind, some of which greatly exceed the severity of the Bhopal accident in 1984.[18]

The public's right to know about health hazards also can be compromised by failure to effectively implement federal and state regulatory statutes. For example, the *Washington Post* reported in 2004 that cities across the country were withholding or manipulating drinking water test results that showed high levels of lead. The result was that many communities were given a false sense of safety of their water supplies. Such data lapses might be discovered with more vigilant federal oversight of the Safe Drinking Water Act, yet the EPA at the time had only 72 enforcement employees for its drinking water program, or one for every 2,238 water systems in the nation. This would seem to be far too few to permit careful checking of drinking water systems nationwide, and the same limitations affect the states (Leonnig, Becker, and Nakamura 2004). Much the same kind of accusations arose in 2009 over the presence of atrazine—an herbicide widely used to protect crops, lawns, and golf courses—in drinking water supplies. It is one of the most common contaminants found in drinking water, but the EPA maintains that the generally low concentrations pose no unusual health risks. Yet its occasional presence at high levels is often omitted from local water system reports and has become a matter of concern to scientists and public health officials. Even many local water officials are unaware that atrazine concentrations sometimes rise sharply in their communities (Duhigg 2009a).

Other examples abound, and are likely to continue in light of severe budgetary constraints facing both the federal government and the states. As discussed earlier, the U.S. GAO reported in 2006, for example, that the EPA was falling short in its efforts to reduce health risks from toxic air pollutants under the Clean Air Act; the agency did not meet some 30 percent of the act's requirements, and "regularly misses deadlines." The EPA was especially likely to be less than diligent in regulating smaller sources of hazardous air pollutants (Heilprin 2006; U.S. GAO 2006). Such findings are not unusual for the EPA, but they do remind us that existing laws and programs often fail to achieve their goals, or do so only after years of delay because of insufficient resources, litigation, or agency priorities that do not target such regulation or its enforcement (Vig and Kraft 2010).[19]

For both regulatory and information disclosure policies to work well, government needs a continuous supply of reliable data on environmental conditions and trends, including those related to toxic chemical use and releases. The need is widely recognized today although disagreement continues over the best way to develop suitable databases and budgetary constraints will limit what can be done at any given time. For example, in 2008, the Bush administration directed federal agencies to develop a set of selected national environmental indicators and they were to be used to facilitate public discourse as well as decision making within agencies. One notable goal of the effort was to provide nationally consistent and more widely accessible indicators following both governmental and private sector initiatives of this kind. The directive was issued after a National Academy of Public Administration report on the subject, and the EPA developed a new Web site for what it calls its Environmental Indicators Gateway (www.epa.gov/indicators/). The TRI is on the EPA's list as an "indicator project," as are such widely used reports such as the agency's Air Trends Online, the vehicle for reporting on air pollution trends, and the National-Scale Air Toxics Assessment (NATA). NATA, which draws data from the TRI and other sources, combines data on toxic emissions, an atmospheric fate and transport model, an exposure model, and the use of health risk criteria to characterize the risks associated with exposure to air toxics. The assessment is based on 177 hazardous or toxic air pollutants, which is a subset of the 187 air toxics regulated under the Clean Air Act, with the addition of diesel particulate matter. As one sign of the limitations of such projects for public information, however, the public version of NATA appears to use data that are about seven years old at the time of publication.[20]

Final Observations

Whether the focus is information disclosure or regulation, a central question about environmental policy in the twenty-first century is whether we have appropriately designed institutions and processes to bring about the outcomes desired. There is little question that the current mix of public policies does not always live up to these expectation, and critics have offered a plethora of recommendations for policy change. We think information disclosure policies such as the TRI program have played a fruitful role in the mix of contemporary policies and regulations. We also conclude, however, that much could be done to improve the program's effectiveness, particularly in ensuring that the information collected is easily understandable and meaningful in terms of real public health risks faced in communities across the nation. Our policy recommendations for changes to the TRI are meant collectively to transform the TRI program to again make it a causal force influencing environmental performance across most facilities. We also think that information disclosure probably works best in combination with the mix of incentives and disincentives provided by other regulatory and nonregulatory programs. Policymakers and analysts sometimes suggest a necessary choice between regulation and voluntary approaches, as though the two were antithetical. We are more persuaded by the logic of a hybrid approach that combines the most positive aspects of the two. The real question ought to be what mix of sticks and carrots is appropriate in a given situation. And what combination makes sense in a federal system where each of the 50 states may well need a slightly different mix.

Some critics of environmental programs, particularly economists and business leaders, point to the need to pay more attention to costs and to the efficiency of program requirements. Certainly the costs that any program imposes on society must receive serious consideration, and policymakers ought to use whatever mechanisms are at their disposal to minimize costs while still achieving the program's goals. Balancing costs and program performance would be easier if more agreement existed on how to measure all appropriate costs and benefits. Programs such as the TRI involve other considerations as well, particularly the public's right to know about certain information and the widely shared expectation for transparency and accountability today in government as well as in the private sector. Proposals to ease the burden of regulation or of information disclosure need to pass a common sense test. If adopted, to what extent might they sacrifice information that is critical to protection of the public's well-being or essential to meet expectations for equity, such as the environmental justice implications noted at the beginning of the chapter?

Finally, we have emphasized the role of social science research in evaluating environmental policies and programs and thus in clarifying how best to achieve policy goals while balancing competing evaluative criteria. We have suggested as well continuing research needs and the kinds of questions that students of environmental policy need to address in the future. Both the importance of the policy questions faced and the complexity of social phenomena being studied reinforce the case for using a multi-pronged research strategy that combines quantitative and qualitative methods. Definitive answers to some questions demand the most rigorous quantitative analysis of available data. Yet other questions can be studied only tentatively and qualitatively through case studies, histories, surveys, and interviews. Each approach has its strengths and limitations, and yet the methods can be integrated in ways that potentially yield more than might be achieved by more narrowly conceived research. We hope that this study of information disclosure and of the TRI program speaks to the need to think about social science research in some new ways that can improve our understanding of public policy and its impacts while simultaneously providing insights that might inform a new round of assessing problems and developing the best policy tools to resolve them.

Notes

Chapter 1

1. For examples of the new corporate attitude toward environmental performance and an increasing preference for greener methods of production, including reduced greenhouse gas emissions, see Fiorino (2006), Laszlo 2003), Esty and Winston (2006), Speth (2008), Gunningham (2009), and Press and Mazmanian (2010), among others.

2. We use the term "information disclosure" throughout the book to denote the release of information to the government and the public, particularly when the release of that information is mandated by law, as is the case for the Toxics Release Inventory. We might differentiate this kind of release of information from what some would term "information provision," where corporations, nonprofit organizations, and governments may make information available to the public, although possibly on a more limited, selective, and voluntary basis. In both cases, we think the important questions are how information is understood and used by the public and others, and what difference the process of collecting and releasing the information makes for environmental decision making.

3. The estimate comes from Portney and Probst (1994), who found that private industry in the early 1990s accounted for about 57 percent of the total cost associated with implementing the seven major environmental protection statutes for which the EPA is responsible. In a later review, Portney (1998) reported that the EPA estimated in 1997 that the annual costs of complying with federal regulations alone for these statutes (that is, not counting tougher state regulations related to the same legislation), was $170 billion. Most of these kinds of estimates are inherently limited. For obvious reasons, it is not easy to estimate compliance costs, and these costs might change significantly over time as industry finds new ways to meet regulatory standards. Portney argued in 1998 that the EPA's estimates likely were too high.

4. These reports from OIRA can be found at www.whitehouse.gov/omb/inforeg_regpol_reports_congress/.

5. One of the best treatments of the general subject of environmental public opinion is Deborah Lynn Guber's *The Grassroots of a Green Revolution: Polling*

American on the Environment (2002) . Similarly helpful in understanding the public's lack of knowledge about environmental issues and policy is Eric R. A. N. Smith's *Energy, the Environment, and Public Opinion* (2002). Despite the many proposals made in recent years to encourage civic environmentalism, public participation in decision making, and information disclosure, analyses like these provide a sobering picture of an inattentive and poorly informed public that is difficult to mobilize on environmental issues. We return to this concern about the public's interest, knowledge, and capacity later in the book as we examine the impact of the Toxics Release Inventory on communities and corporations. See also Atlas (2007) on how much the general public knows about and cares about the TRI information, and how often the public contacts industrial facilities about their TRI releases.

6. As we stated in note 2, not all information disclosure policies are mandatory as is the TRI program and the newly created federal greenhouse gas release reporting requirement. Any effort to compare such policies systematically across many substantive areas of concern would benefit from a typology that distinguishes those that are mandatory from those that are voluntary, along with other significant characteristics. Mandatory policies involve some degree of oversight by government agencies and clear specification of what kind of information is to be provided and in what format, whereas many voluntary efforts at information release are less structured and may involve no third-party auditing of any kind, thus to some extent undermining their value to the public.

7. The press release by the coalition is cited in Baue (2008).

8. The new EPA greenhouse gas reporting rule is available at: http://epa.gov/climatechange/emissions/ghgrulemaking.html. Under the new requirement, the first reports would be due at the end of March 2011, reflecting emissions for the 2010 calendar year. All facilities with annual emissions of 25,000 megatons or more of CO_2 equivalent must report to the EPA, and the reports are to be publicly posted. Depending on how Congress acts on pending climate change legislation, the agency may move to more of a regulatory stance on greenhouse gas emissions, which may then link the new reporting requirement with a formal regulatory program. The agency announced in 2009 in a final "endangerment" rule that greenhouse gases posed a threat to human health and the environment. Should Congress not set climate change policy, that decision will allow the EPA to develop emissions limits not only for industry but for vehicles and other major sources of greenhouse gases. See Broder (2009).

9. The statement comes from the project's Web page at Harvard University: www.ksg.harvard.edu/taubmancenter/transparency/index.htm.

10. Dow Chemical Company bought Union Carbide in 2001, although questions remain about whether Dow assumed responsibility for the company's liabilities, including obligations to pay for cleanup of the site. Dow denies that it has any responsibility for a site cleanup. In 1989 it paid $470 million in compensation to the Indian government, with the understanding that the government would be responsible for site cleanup.

11. As we discuss in chapter 7, the EPA is now committed to releasing the annual TRI reports in a more timely manner than has been the case, partly in response

to criticism about dated information. The historical two-year delay should disappear, as the agency states strives to release the data in the same calendar year that they are received. It kept to that schedule during 2009, releasing the 2008 TRI data in early December of that year.

12. A 2006 report by the U.S. Government Accountability Office (GAO) faulted the EPA for not making more progress on the implementation of the air toxics requirements of the 1990 law. The GAO said that most of the EPA's actions were completed late and many significant portions of the program had yet to be addressed 16 years after adoption of the act. While the EPA has made some progress in regulating large stationary sources, it has achieved much less with small stationary and mobile sources, which account for most of the air toxics emissions. Moreover, the GAO asserted that the EPA had no comprehensive strategies for implementing the act's unmet requirements, nor even very good data on the degree of risk reduction that has been achieved through its regulation of air toxics. See U.S. GAO (2006).

13. The study of these facilities was conducted by the Center for American Progress, a liberal Washington think tank.

14. Michael Chertoff, secretary of the Department of Homeland Security from 2005 to 2009, and the Bush administration itself concluded that voluntary efforts by industry would be insufficient to achieve security goals. Only a small number of the 15,000 facilities nationwide that use large amounts of hazardous chemicals participated in a voluntary self-assessment program initiated by the American Chemistry Council, the leading trade association for chemical manufacturers.

15. See an editorial in the paper, "Smoggy Numbers: Wide Variances in Estimates of Industrial Plant Toxic Emissions Challenge Reporting System's Credibility," *Houston Chronicle*, May 11, 2006, B8. For a more thorough critique of the reliability of TRI data, see reports of the Environmental Integrity Project. The group asserted in a 2004 report, "Who's Counting: The Systematic Underreporting of Toxic Air Emissions," that the TRI database seriously underreports releases of air toxics from chemical plants and refineries. Reports are available at the group's Web site: www.environmentalintegrity.org/.

16. In January 2007, the EPA finalized its rule for the TRI's burden reduction goal by substantially raising the threshold for what facilities can release without having to disclose the details on the agency's Form R. However, as we report in chapter 7, in early 2009 Congress reversed the Bush administration rule, an action that the Obama administration applauded. One of the concerns raised was the impact that the Bush rule would have on environmental justice. An influential GAO report (2007b) concluded that many of the TRI facilities that no longer would have been required to submit Form R were located in minority and low-income communities; hence the loss of information available to the public could have affected those communities disproportionately.

Chapter 2

1. The return rate for the corporate survey was 24 percent, equal to or better than comparable surveys of business firms, thus providing confidence in the

survey results. For the public officials survey, the rate varied from 38 percent for local officials to 80 percent for federal officials. We discuss the survey administration and responses in detail in chapter 5.

2. We discuss the RSEI model and other aspects of our measures of environmental performance later in the chapter but in greater detail in chapter 3.

3. Information disclosure needs to be considered as an influence on release reductions, but it is also important to keep in mind that other factors such as production changes at facilities can have an influence. Similarly, political dynamics at the local level could make a difference, as we will discuss in chapter 4.

4. Their dilemmas are, in turn, variations on the classic Prisoner's Dilemma. The Prisoner's Dilemma (Rapoport and Chammah 1965) lays out a stylized decision game where two players have the ability to influence the other's payoff depending on the decision they each make. The assumption is made that neither player can bind the choice of the other; choices are made independently. The game is structured in such a way that players work to avoid their worst payoff (in this case, 1 point), thereby choosing a suboptimal payoff (2 points) even though there is the potential for a better outcome (3 points) through cooperation.

5. There are strong parallels here between our performance dilemma and the collective action dilemma presented by Press and Mazmanian (2010). They extend Potoski and Prakash (2004) in their discussion of the greening of industry and the potential conflict between governmental action (command-and-control regulation versus flexible regulation) and corporate action (evasion versus self-policing). We take another step along this theoretical path to highlight the environmental performance of the facilities.

6. The relationship between governments and facilities is embedded in a broader context of relationships with legislatures, interest groups, and citizens. As Scholz (1991) described it, the enforcement dilemma is embedded in nests of principal-agent games.

7. Fiorino's arguments about policy learning and policy change are in the tradition of work by Paul Sabatier and Hank Jenkins-Smith (1993) on these subjects and similar to the insights offered by Sabatier's advocacy coalition framework, the most recent version of which can be found in Sabatier (2007). As we discuss in chapters 5 and 6, corporate experience with the TRI reporting process clearly has affected the behavior of many facility officials, and experience in working with corporations also has affected the views of federal and state regulatory officials. We would expect that over time, these experiences would alter the way an information disclosure requirement is perceived as well as the relationships between corporate and government officials.

Chapter 3

1. In some content analyses of news coverage of the TRI between 1991 and 2000 (which we do not report on here), we found a clear drop off in coverage nationwide.

2. This number includes chemicals added to the list in the years after the beginning of the program.

3. See Duhigg (2009b, 2009c). The articles offer unusually long and detailed analysis of contamination of both surface and drinking water nationwide, complete with online and interactive graphics to assist readers in learning more about local water problems, incidents of noncompliance with the Clean Water Act and Safe Drinking Water Act, and enforcement actions across the country.

4. For an extended discussion of how the Kingdon model of agenda setting applies to environmental policymaking, see Kraft 2011 (chapter 3).

5. One interesting source of data for this period comes from an environmental advocacy organization, the Environmental Working Group. It has a project called Chemical Industry Archives that reports on industry documents made available during litigation, including the industry's response to the Bhopal disaster. See www.chemicalindustryarchives.org/dirtysecrets/Bhopal.index.asp.

6. It is worth noting that the politics of the mid-1980s was more conducive to information disclosure policy than it was to a stronger regulatory revision. The Reagan administration had argued strongly against new regulatory powers for federal agencies and information disclosure would have been a less bitter pill to swallow. See Vig and Kraft (1984) for an assessment of the administration's environmental policy rhetoric and actions.

7. The quotation is taken from the agency's description of the program, "What Is the Toxics Release Inventory (TRI) Program?" It is available at: www.epa.gov/tri/triprogram/whatis.htm.

8. The various reporting requirements are detailed on the agency's TRI Web pages and are discussed at length in the appropriate sections of the Code of Federal Regulations.

9. Facilities exempt from reporting are contributors to toxic emissions, and their relative impact in comparison to facilities that do report to the TRI are thought by some (e.g., Sheiman 1991) to have the potential to be quite significant. A wider discussion of environmental impacts would have to also include an analysis of non-point sources of pollution, which is well beyond the scope of this book.

10. Sigman (2000, 244) notes that under the Toxic Substances Control Act (TSCA), the U.S. EPA compiled an inventory of some 62,000 chemicals that were produced or imported when TSCA was enacted (1976), with perhaps another 10,000 added between the mid-1970s and 2000. In a recent review of TSCA's limitations for regulating toxic chemicals, Schwarzman and Wilson (2009) note that some 34 million metric tons of chemicals were produced in, or imported into, the United States *every day*, and over the next 25 years global chemical production is expected to double. Despite this enormous quantity of chemicals, they note that "most health and ecological risks associated with industrial chemicals are still poorly understood," in part because TSCA does not require manufacturers to produce basic information on the chemicals' use, health effects, or likely exposures. Neither do similar policies in other nations. By one recent count, for example, about 17,000, or nearly 20 percent, of the chemicals thought to be used in commercial quantities by 2010, are protected as trade secrets under

TSCA. The names of the chemicals and their physical properties are kept from consumers and nearly all public officials under a provision of the law intended to protect trade secrets in a competitive chemical industry; the chemicals are known to a small number of EPA staff who are barred by law from sharing the information with other public officials, including state health personnel and emergency responders. About 150 of these chemicals are made in amounts greater than 1 million tons a year. The Obama administration has signaled an interest in changing such provisions in the law (see Layton 2010).

11. The numbers come from the EPA's Web site for the TRI program.

12. Although a detailed analysis of enforcement data is beyond the scope of this project, such an analysis is clearly worthy of scholarly attention, and we hope that others take up the challenge. As enforcement databases improve in sophistication, collection and analysis of the data should become easier.

13. Due to the nature of our data, the complexity of categories is higher than this illustrative matrix suggests. To be precise, because we have three years worth of data, we have cases where facilities had different patterns of behaviors between 1991 and 1995, and then again between 1995 and 2000. For example, we have facilities that decreased their emissions and risk levels from 1991 to 1995, but then increased both between 1995 and 2000. What this means in effect is that though we have the pure examples of green facilities and brown facilities, we have numerous examples in between. In general, for expository reasons, blue facilities are those that see risk improvements without seeing emission improvements. For yellow facilities, it is the opposite—emissions improvements without risk improvements.

14. The color coding we use is similar in very broad ways to the sort of performance ratings used by Indonesia in the last decade (World Bank 2000) and China in more recent years (Wang et al. 2004). Such ratings allow for the straightforward transfer of information about environmental performance without being reduced to simplistic description.

15. Some may observe that because we use directional categories instead of actual volume changes, our sample may mischaracterize industrial environmental performance as facilities move production abroad or shift emissions to other media besides air. First, because our sample includes facilities reporting in 1991, 1995, and 2000, we do not include polluters that moved off-shore or polluted below the regulatory threshold in our time frame. The TRI requires a threshold of 10,000 lbs of toxic waste in a facility's production related waste, so our sample never captured facilities that found regulatory havens in other countries. Second, pollution volume trends in our survey sample also suggest our analysis is not capturing a large number of facilities with unusual or even fictitious decreases. In 1991, 12 percent of our sample reported releases below 500 pounds. In 2000, this rose by 10 percent as 22 percent of our sample dropped below 500 pounds. Of those, only 10 had 1991 pollution volumes between 1,000 and 10,000 pounds, 8 between 10,000 and 100,000 pounds, and only 1 with more than 100,000 pounds. In sum, only 19 facilities dropped emissions exponentially—8 percent of our sample. Likewise, 18 percent, or 45 facilities, of our survey sample in 1991 reported releases greater than 100,000 pounds. Only three of those

reported 2000 releases below 1,000 pounds and 4 dropped between 1,000 and 10,000 pounds.

Chapter 4

1. The descriptions and analyses in chapters 3, 5, and 6 instead use a longer time frame that covers 1991, 1995, and 2000.

2. We relied on the four Census categories of state regions. The Northeast region (State code 91) includes Maine, New Hampshire, Vermont, Massachusetts, Rhode Island, Connecticut, New York, New Jersey, and Pennsylvania. The Midwest region (State code 92) includes Ohio, Indiana, Illinois, Michigan, Wisconsin, Minnesota, Iowa, Missouri, North Dakota, South Dakota, Nebraska, and Kansas. The South region (State code 93) includes Delaware, Maryland, District of Columbia, Virginia, West Virginia, North Carolina, South Carolina, Georgia, Florida, Kentucky, Tennessee, Alabama, Mississippi, Arkansas, Louisiana, Oklahoma, and Texas. The West region (State code 94) includes Montana, Idaho, Wyoming, Colorado, New Mexico, Arizona, Utah, Nevada, Washington, Oregon, California, Alaska, and Hawaii.

3. In one of the seminal analyses of comparative environmental policy efforts (Lester 1994), states were characterized as progressive, struggling, delaying, or regressive based on a typology of commitments to environmental protection (high and low) and institutional capacity (strong and weak). We follow this descriptive approach, but instead use quartiles, which allow us to evenly divide states into four categories of industrial environmental performance, two above and two below the mean.

4. Content validity refers to the dimension of measurement involving the extent to which a variable covers the breadth of a concept. As Carmines and Zeller (1979) explained, a measure of mathematical knowledge would be weak in content validity if it captured only addition and subtraction and not multiplication and division. Comparative measures of industrial environmental performance result in the same problem by omitting risk content, as discussed in chapter 3.

5. Our subsequent modeling presumes that variables are related to one another in a linear fashion. This is one of the several assumptions required for linear regression analysis to obtain the "best linear unbiased estimates" (Lewis-Beck 1980). Correlation is a statistical technique that fits a straight line on a scatter plot of two variables (see figures 4.1 and 4.2). For instance, previous research has found that more politically liberal states tend to spend more on environmental regulation, so we expect these two variables to produce a Pearson's coefficient between 0 and 1 (Johnson and Reynolds 2005).

6. Since our correlation and regression analysis assumes linear distributions among variables, the highly skewed pollution measure required transformation into a smoother distribution. According to Hamilton, "Regression requires no assumptions about the distribution of [independent] variables, but in practice skewed distributions are often associated with statistical problems such as

influence and heteroscedasticity" (Hamilton 1992, 55). Logarithmic transformations are a common technique to modify skewed variables.

7. While air pollution program strength might seem like the more appropriate measure, state efforts in this area typically focus on pollutants like sulfur dioxide, nitrogen oxides, particulates, and other airborne compounds regulated under the National Ambient Air Quality Standards (NAAQs) of the federal Clean Air Act. In the TRI program, facilities are required to report on the fate of components in their Production Related Waste (PRW), including releases, recycling, treatment, and energy recovery of over 300 toxic chemicals unregulated by the Clean Air Act (CAA). In the hierarchy of the EPA's waste management preferences, source reduction is the most preferred, followed by recycling, energy recovery, and treatment. Disposal or release into the environment is the least preferred, or the management choice of last resort. Thus, state effort in regulating hazardous waste is the more appropriate variable because it measures a state's commitment to cleaning up industrial pollution before it is released into the environment. Positive and statistically significant correlations occurred between hazardous waste expenditures and all three industrial environmental performance measures. States that spent more on regulating waste before it was released into the environment had industry that reduced more air releases and relative risk.

8. The variable contained six components. States received a 1 if they had legislation that required or promoted pollution prevention planning by industry for the first component. A second component (scored a 1 or 0) involved some other pollution prevention activity such as multimedia coordination, database integration, pollution prevention training, or pollution prevention technical assistance.

In the remaining four components of this index, the EPA categorized pollution prevention efforts in three ways. A well-established program scored a 1, a partially completed effort scored a 0.666, and an effort that was just initiated or planned received a 0.333. In addition to legislation, states could initiate multimedia approaches in either inspections or enforcements, making up two more components of our index. Many states began multimedia efforts in their inspections and enforcement because, according to the EPA (1993, 3), emission reductions could be gained when public agents looked more holistically at facility waste rather than just by air, water, or solids.

Likewise, the other two index components included efforts to integrate pollution prevention into state inspections or enforcements. For inspections, scores were achieved if some form of pollution prevention technology transfer or technical assistance referrals were being made by state environmental inspectors. On the enforcement side, several states began to require pollution prevention measures or waste-reduction plans in their compliance actions against facilities.

9. The questions asked if: (1) the state made TRI data available to the public; (2) health or risk information was provided; (3) TRI data were geographically analyzed; (4) the state conducted environmental equity analysis; (5) a state computerized TRI data; (6) the public could access the state database; (7) the state customized database reports; (8) libraries received state TRI data; (9) LEPCs received state TRI data; and (10) other agencies or organizations provided the public with computerized TRI information.

10. Although this finding might seem to be counterintuitive, it probably reflects the rather simple measures that have been used to describe state information disclosure programs. Studies have measured only whether a state has or does not have a program, and not what the state is actually doing in pollution prevention. As a result, the message that is being sent to industry is not captured by this measure. Even the EPA assessment of state information disclosure programs is based on a rather crude measure of state activity. It distinguishes between highly developed programs and nascent ones, measuring state efforts to disseminate more information but not efforts to modify the information. Yet in pollution prevention programs, the more successful states are engaged in other activities as well, such as offering technical assistance to industry. What is likely to be most important to industry is whether state activities on information disclosure actually alter the performance dilemma as seen by industry. We would expect industry to respond to genuine efforts by government to help decrease pollution, such as through well-designed and supported programs of pollution prevention and technical assistance, particularly when offered through a collaborative relationship with industry.

11. Interval and ratio measures contain numerical information as opposed to the qualitative information in ordinal and nominal variables. Interval measures mean that the distance between values, or interval, is meaningful. However, interval measures do not have a true zero point like ratio measures.

12. R-squared is the common statistic used to describe how a regression model fits the data points, or a measure of the spread of data points around the regression line. King is an outspoken critic of the use of R-squared, but does identify one useful application. "You can directly apply and evaluate R^2 when comparing two equations with different explanatory variables and identical dependent variables. The measure is, in this case, a convenient goodness-of-fit statistic, providing a rough way to assess model specification and sensitivity" (King 1986, 677).

13. We first looked for evidence of multicollinearity among our independent variables, a common problem when the independent variables are highly correlated in a multiple regression and can generate problems in the model's parameter estimates (Lewis-Beck 1980, 58). We calculated the Variance Inflation Factors (VIF) for different combinations of independent variables and avoided combinations where a VIF exceeded four. VIF is a technique to identify collinearity among independent variables that multiple regression estimates. When independent variables are highly correlated, standard errors of fitted coefficients tend to be high. The VIF procedure provides a measure of the multiple correlation coefficient when one candidate independent variable is regressed against all other candidate independent variables. The VIF formula is $VIF(xj) = 1/1\text{-}R2j$. If one of the independent measures (xj) is linearly related to other explanatory variables in a significant manner then R2j would be close to one and the VIF quantity would be large. The remaining diagnostics involve the error term, or the last term in the standard regression equation: $Y_i = a + bX_i + e_i$. We examined scatterplots of the prediction errors or residuals of our regression models. First, we looked for evidence of outliers and found none. An outlying state with an extreme value in our measures could skew the multiple regression results. Second, we confirmed that there was no evidence of heteroskedasticity in the residuals,

or an increase in the prediction errors as the value of X increases. Third, we confirmed that our residuals were also normally distributed (see Lewis-Beck 1980, 26).

14. The path model included the following four regression equations:

$X4 = a + bX1 + bX2 + bX3 + e$

where X4 is a state's level of toxic waste regulation expenditures, X1 is a state's median household income, X2 is a state's pollution severity, and X3 is a state's public liberalism; b refers to the standardized regression coefficients, and e is the standard error.

$X6 = a + bX1 + bX3 + e$

where X6 is a state's environmental groups strength, X1 is a state's median household income, and X3 is a state's public liberalism.

$X8 = a + bX2 + bX4 + bX5 + bX7 + e$

where X2 is a state's pollution severity, X4 is a state's level of toxic waste regulation expenditures, X5 is a state's administrative professionalism, and X7 is a state's industry group strength.

$Y1 = a + bX6 + bX7 + bX8 + bX9 + e$

where Y1 is a state's share of pollution reducing or performing facilities, X6 is a state's environmental group strength, X7 is a state's industry group strength, X8 is a state's pollution prevention integration, and X9 is a state's growth in manufacturing production worker hours.

15. We used *Clarify* to simulate a thousand Americas modeled after New York, Georgia, and the other thirty-nine states in our sample with the constellation of variables under analysis. The difference in state industrial environmental performance between states with the most and fewest environmentalists amounted to a 2 percent increase in the frequency of TRI facilities getting safer and cleaner. This difference is measured while holding industry influence, pollution prevention efforts, and manufacturing growth at their means. That amounts to five more facilities recording reductions in air pollution emissions and risk in a state with the industrialization of New York. Likewise, in the states leading in manufacturing growth, the frequency of safer and cleaner facilities was 3 percent less than those states where industry production shrank. A state with the most growth in manufacturing would see nearly eight fewer facilities achieving lower emissions and risk while its environmentalism, policy effort, and industrial influence were average. Conversely, in a state where industry was a dominant economic force or where environmental regulators led all others in pollution prevention integration, each condition meant eight safer and cleaner facilities when the other three independent variables in our model were held at their mean. Based on the assessment we provided in chapter 3, the average green facility decreased 252,937 thousand pounds of air pollution. Eight of these facilities would translate into over two million pounds of toxic emission reductions.

16. Rabe (1986) first examined the early and incremental attempts by Illinois, New York, and Washington to move beyond fragmented environmental policies and toward integrated efforts. He continued to investigate integrated environ-

mental management cases in New Jersey (Rabe 1991), Minnesota and New Jersey (Rabe 1995), and the Great Lakes (Rabe 1996; Rabe and Gaden 2009). Yet pollution prevention came into its sharpest relief in a comparison of four states and four Canadian provinces through the lens of the principal-agent theory. Drawing from this perspective, Rabe made this observation: "Prevailing views on environmental policy would lead to the hypothesis that decentralization from national to subnational units and delegation from legislative principals to bureaucratic agents would unleash creative energies and policy innovation" (Rabe 1999, 304). Our results show that states did not uniformly support this innovation expectation, but some did. In those leading states, pollution prevention leadership was a function of bureaucratic capacity, which in turn was a function of political dynamics.

Chapter 5

1. The survey was developed after consultation with other scholars who have used such instruments. These include Dorothy Daley, Magali Delmas and Dennis Aigner, Mark Lubell and John Scholz, and Paul Sabatier. We are grateful to all of them for their willingness to share questionnaires and interview schedules with us. In the end, we used few of the same questions that had been asked in these surveys simply because our focus was so different. However, the format of our questionnaires did draw from the style and arrangements that many of our colleagues had found useful. The final questionnaires that we used are available upon request. As we noted in the preface, we have archived all survey results through the Data Preservation Alliance for the Social Sciences project of the Inter-University Consortium for Political and Social Research at the University of Michigan.

2. In an e-mail to interested scholars in 2004, Matthew Clark of the EPA's National Center for Environmental Research and Quality Assurance reported that response rates for facility-level surveys of multiple industry sectors were typically quite low, from 7 percent to 27 percent, largely because there is no necessity for the facilities to respond and doing so both consumes valuable time and potentially releases information that the facility would prefer to keep private. See our methodological appendix in this chapter for a discussion of how we tried to compensate for these conditions in the way we framed the cover letter and designed the survey instrument.

Scholars can compensate for the lower than average response rates with different sampling techniques and stratification approaches. In our case, the response rate was not only at the high end of this range, but as we note in the text of the chapter, our comparisons of the respondents, the sample to which the surveys were sent, and the general sample of TRI facilities that we report on in this volume confirmed that we did indeed have a representative sample.

3. Almost all of these 8,476 facilities also reported in 1991. More importantly, all of our eventual respondents also reported in 1991.

4. EPA doubled the reportable chemical list in 1996, thus potentially distorting longitudinal analyses.

5. A separate analysis was done comparing the latitude and longitude of survey facilities with the larger sample of facilities. The regional distribution matched up well. That analysis is available from the authors upon request.

6. We calculate the response rate based on the number of questionnaires returned or online responses recorded (N = 237) in comparison to the number of surveys that we sent out for which we could confirm delivery. We mailed 1,083 questionnaires to our sample, but of these only 979 were actually received by the recipient. The difference reflects industrial facilities that had closed, were acquired by another company, or moved after their last TRI report (either within the United States or overseas), and for which we could not find a current address. An added complication in this kind of survey work is that the EPA database is not as usable as one might assume. Facility addresses, for example, often represented the physical location of the facility rather than a mailing address. The U.S. Postal Service returned a rather large number of surveys to us (about 18 percent of those mailed), and we then had to find the actual mailing address. We were successful in most cases but not all. For the benefit of others who attempt such surveys, we found that telephone calls to the TRI public contact official at the facilities frequently were not returned.

In what may be a valuable lesson based on this experience, we concluded that manufacturing facilities have more of a dynamic character than is often assumed in discussions of the TRI or of pollution control more generally. Indeed, one respondent called to "vent" that his company no longer filed TRI reports because "all of our manufacturing operations have been shifted to Mexico, India, and other countries." We couldn't help but wonder about how much of the reduction over time in reported TRI releases might be attributable to such declines in the manufacturing presence in various localities in comparison to actual improvements in facility environmental performance. We had no way to test directly for such shifts, but the topic clearly merits future study.

7. To be precise, three of the interviews were performed at the pilot stage of the project. The rest occurred after the survey sample was chosen.

8. There was also some initial hope we might be able to link county-level data to the different groups of facilities, but these analyses were dropped due to the low number of interviews in several of the counties.

9. The experience of public officials in their current positions also varied significantly, with respondents ranging from six months in their current position to 30 years (In some cases local respondents may have been talking about their experience in their work setting rather than with the LEPC). Respondents spent an average of roughly nine years in their current position. Public officials came from a wide set of academic backgrounds, although almost half had at least a bachelor's degree in the physical sciences, biological sciences, or engineering. LEPC respondents tended to be evenly distributed across multiple education categories, ranging from a high school diploma to an advanced degree. State employees and federal employees tended to have higher levels of education.

10. In some cases respondents may have given the number of employees at their firm instead of at their specific facility.

11. As noted earlier, the data results are available from the Inter-University Consortium for Political and Social Research. The online survey is no longer available but it essentially duplicated the survey instrument in a Web-based format, permitting answers to be recorded directly into a database that was later converted to Excel spreadsheets. Our results were made available to all respondents. Many of them took advantage of the offer and told us they found the results to be quite interesting.

Chapter 6

1. The differences in industrial sectors deserve further scrutiny. Responsible Care may be the strongest explanatory factor for the chemical industry's release reductions, but our current analyses can neither prove nor disprove this argument.

2. It is worth keeping in mind that we are not controlling for production levels in comparing industrial sectors. The plastics industry may have more brown facilities simply because production levels have increased. Our use of "brown" as a category does not account for the environmental efficiency of a facility. Further study would need to occur to better understand which industries are doing a better job of reducing toxic releases per unit of production.

3. In this case, all mixed facilities are excluded from the analyses.

4. Often EMSs require that a facility be in contact with LEPCs.

5. The most interesting results come from the comparison of green and brown facilities exclusively. Comparing greens to all other facilities shows no significant variation, while comparing browns to all others results in findings similar to those mentioned here.

6. These comments are partly based on some correlation analyses not reported here, but available from the authors upon request.

Chapter 7

1. As we noted in chapter 1, the EPA burden reduction rule was to raise the threshold (essentially from 500 pounds a year to 2,000 pounds) for what facilities can release without having to disclose the details on the agency's Form R. Under the new rule, some 6,500 facilities could have converted to the shorter Form A, and the EPA estimated that each would save about 25 hours of reporting time in making the change. See Skrzycki 2006, D1.

2. The GAO said in particular that facilities that no longer would submit Forms R tended to be located in minority or low-income communities, and that the "reduction in toxic chemical information could disproportionately affect them." In addition, a nonprofit group, the U.S. Public Information Research Group, said its analysis of the proposed rule changes indicated that 922 of the nation's 33,000 residential zip codes would lose all of the detailed pollution data they now get if companies chose to use the shorter Form A.

3. Among other comments the SAB said that the TRI data "provide the only reliable source of longitudinal data for this type of research" (spatial analyses of toxic hazards) and noted that over 120 scholarly articles had been published using the TRI data to address "a wide range of public health, economic, and social science issues" (Ascher, Steelman, and Healy 2010).

4. The quotations from Nelson appear in Skrzycki (2006). On the state lawsuit, see DePalma (2007, A16). The GAO report was given in testimony before Congress. See U.S. GAO (2007a).

To put concerns over TRI burdens on industry into perspective, the federal Office of Management and Budget reports that the Paperwork Reduction Act of 1980 has done little to slow a rising volume of federal forms and the time needed to complete them. It estimates that by 2009 Americans spent some 10 billion hours filling out more than 8,000 different federal forms and various requests for information, up from 1 billion hours in 1981. Even with vastly improved tools for sorting through this massive amount of data, such a surfeit of information potentially can become unmanageable and undercut the utility of collecting it in the first place. It is less clear how the TRI burden compares to other paperwork requirements for business, and our surveys of corporate officials provides a mixed picture on the question; it has been a burden to some and for many it is now less of a burden, primarily because of online reporting options. See Cowan 2009.

5. The quotations are taken from announcements on the EPA's Web page devoted to the greenhouse gas rule, and from its news release of September 22, 2009. See www.epa.gov/climatechange/emissions/ghgrulemaking.html.

6. Although the book focuses on national, state, and local developments in the United States, there is comparable interest internationally in promoting information disclosure as part of a broader effort to foster the greening of industry and sustainable development. For example, the Canadian National Pollutant Release Inventory system is similar to the U.S. TRI program, although Canada has not yet linked industrial releases with estimates of health risks as the U.S. program has with its RSEI model. For global actions, see the work of the World Business Council for Sustainable Development; a new Global Reporting Initiative, which provides guidance for sustainability reporting (www.globalreporting.org/Home); and the environmental work of the Organization for Economic Co-operation and Development. Borck and Coglianese's (2009) review of the state of knowledge about voluntary environmental programs indicates some of the characteristics that any comparative or cross-national study of such programs ought to examine.

7. The report is discussed in Gilbert (2008). It is called "The Smokestack Effect: Toxic Air and America's Schools," and it is available at http://content.usatoday.com/news/nation/environment/smokestack/index.

In the case of Wisconsin public schools, the state Department of Natural Resources (DNR) spent considerable time reviewing the study's findings and concluded that the methodology was substantially flawed and that the study needlessly alarmed parents and other citizens in the state. The DNR preferred to use a far more sophisticated, modified version of a regional air impact modeling initiative developed by staff in the Region 6 office in Dallas that covers both stationary and mobile sources of air releases. They found the RSEI model to be too limited to reliably estimate public health risks of air releases, and they recognized a continuing need to figure out how best to display the information to reduce the chance of misinterpretation (personal communication from a DNR toxic chemical specialist, April 15, 2009).

8. The quotations are taken from an article on the Environmental Defense Fund Web site that recounts the history of the Scorecard effort: www.edf.org/article.cfm?contentID=4940. The account is reprinted on the Scorecard Web site.

9. The site is www.epa.gov/oppt/rsei/pubs/get_rsei.html.

10. The report, now called the TRI National Analysis rather than the TRI Public Data Release, is available at the EPA's Web site: www.epa.gov/TRI/tridata/tri08/national_analysis/index.htm.

11. Mark Atlas suggests that the National Emissions Inventory could potentially provide the public with much more useful data and in effect replace the need for the TRI program. The money saved from replacing the TRI could, he suggests, pay for thousands of government inspectors of regulated facilities. The facilities might resist such a change, but the public would benefit through a more inclusive and more accurate data collection process (personal communication from Atlas, June 8, 2007).

12. For a discussion of how Twitter can be used for environmental communication, see Clark (2009). Twitter is already being used for environmental organizing such as grassroots lobbying campaigns. It is also used as a forum by environmental professionals. Clark reviews and cites a number of articles and briefing papers on how Twitter and other social network sites might be used to work on environmental problems. What has been done to date might be supplemented by use of Web 2.0 and other social media to assist in informing citizens of new data and developments and in improving communication between citizens and both companies and government. For a review of how wikis might be used to improve civic participation in general, see a review of a new book, *Wiki Government*, by Shneiderman (2009).

13. In late 2004, the EPA's Office of Pollution Prevention and Toxic Substances released a report called *Community Air Screening How-To Manual, A Step-by-Step Guide to Using Risk-Based Screening to Identify Priorities for Improving Outdoor Air Quality*. The manual illustrates how government agencies might take action to improve the public's capacity to understand and use data, in this case TRI data. Indeed, it is seen as an integral part of the agency's Community Action for a Renewed Environment (CARE) program, which was begun in the fall of 2004, and continues to be funded through a competitive grant program designed to reduce toxic pollution in its local environment. About a dozen community projects are funded each year. See the program's description at: www.epa.gov/CARE/index.htm. The manual released in 2004 strongly endorsed the establishment of partnerships between communities and local industry, with broad stakeholder involvement, as the best way to establish local priorities and promote their achievement. The agency's educational effort focused on the RSEI model that we use in the book.

14. For the national office, see www.epa.gov/tri/index.htm. For a typical regional office, see the Web page for Region 10: http://yosemite.epa.gov/R10/OWCM .NSF/tri/trihomepage.

15. In one case, the current contact at a facility asked if we were using an old database because the individual to whom we had written had left the position six years earlier. Yet we were using the most current EPA contact information at the time.

16. The site is http://chemicalright2know.org/, and as of December 2009 it included reports on new TRI developments, resources available for those interested in the program, a listing of recent research and analysis related to the program, examples of the use of TRI data, and an array of user commentaries.

17. A description of Green Tier can be found at the Wisconsin Department of Natural Resources Web site: www.dnr.state.wi.us/org/caer/cea/environmental/. The program is described as a way to provide "credible, creative ways to enable your business to be a powerful, sustainable force for environmental good and enhance your productivity, cut your costs and strengthen the health of your culture and community." The program is designed to "to help environmentally responsible companies achieve environmental and economic gains," through a system of collaborative contracts and charters that are developed jointly by the participating businesses and the DNR. The idea is to streamline environmental requirements and encourage the use of new environmental technologies.

18. The assessment comes from an editorial in the *New York Times*, "Inside the Kill Zone," May 22, 2005, 11. The need for such access is reinforced by recent attempts by some companies to withhold information about chemical accidents, even where doing so constrains the work of emergency responders. For example, in 2008, a large explosion occurred at a West Virginia chemical plant owned by Bayer CropScience. Under the guise of protecting national security, the plant managers refused for hours to disclose the nature of the explosion or the chemical it released. That accident killed two workers and sickened six of the firefighters who arrived at the scene, but it could have been far worse had a tank holding up to 40,000 pounds of methyl isocyanate (the same chemical responsible for deaths and injuries in Bhopal in 1986) been struck. That tank was located just 50 to 75 feet from the explosion site. Investigations later showed that the company often had classified documents as being "security sensitive" when they were not, and that it had a strategy to "marginalize" a local citizen group and the local newspaper in an area of the state that has been long known as Chemical Valley. Bayer CropScience announced in August 2009 that it would substantially cut production of methyl isocyanate at the plant because "we have taken seriously the concerns of public officials and site's neighbors." See Wald (2009b) and Hamill (2009a, 2009b).

19. One constraint on implementation and presumably on corporate behavior can be seen in the Bush administration. It quietly shifted from a strategy of prosecuting corporations to one of deferred prosecution agreements where the government collected fines and appointed an outside monitor to oversee corporate reforms, but without going through a trial. The details of these agreements often were kept secret, making it hard to determine what effect, if any, there was

on corporate environmental performance. From a policy perspective, analysts might ask whether the probability that a company would engage in illegal behavior increases if it believes that indictment, trial, and conviction are unlikely. See Lichtblau (2008). As noted in chapter 2, a major study by the *New York Times* released in September 2009 (Duhigg 2009b) strongly suggested that lax enforcement of the Clean Water Act by the states left millions of citizens with potentially unsafe drinking water. As applied to information disclosure policies and their effects, if companies are less likely to face stringent regulation and significant penalties for noncompliance, they might also not take the TRI data collection and dissemination as seriously as they would otherwise.

20. This is according to the description of NATA on the EPA Web page for the program: www.epa.gov/ttn/atw/nata1999/index.html. The 2002 national-scale assessment based on the 2002 emissions inventory was released to the public in June 2009.

References

Abel, Troy D. 2008. Skewed Riskscapes and Environmental Injustice: A Case Study of Metropolitan St. Louis. *Environmental Management* 42 (2):232–248.

Abel, Troy D., and Mark Stephan. 2000. The Limits of Civic Environmentalism. *American Behavioral Scientist* 44 (4):614–628.

Abel, Troy D., Mark Stephan, and Michael E. Kraft. 2007. Environmental Information Disclosure and Risk Reduction among the States. *State and Local Government Review* 39 (3):153–165.

Agyeman, Julian, and Briony Angus. 2003. The Role of Civic Environmentalism in the Pursuit of Sustainable Communities. *Journal of Environmental Planning and Management* 46 (3):345–363.

Anderson, James E. 2006. *Public Policymaking: An Introduction*. 6th ed. Boston: Houghton Mifflin.

Andrews Richard N. L. 2003. *Environmental Management Systems: Do They Improve Performance?* Chapel Hill: University of North Carolina, National Database on Environmental Management Systems Final Project Report: Executive Summary, January 30.

Andrews, Richard N. L. 2004. "Formalized Environmental Management Procedures: What Drives Performance Improvements? Evidence from Four U.S. Industries." Paper presented at an EPA-sponsored workshop on Corporate Environmental Behavior and the Effectiveness of Government Interventions, Washington, D.C.: April.

Andrews, Richard N. L. 2006. Risk-Based Decisionmaking: Policy, Science, and Politics. In *Environmental Policy*. 6th ed., ed. Norman J. Vig and Michael E. Kraft. Washington, D.C.: CQ Press, 215–238.

Ascher, William, Toddi Steelman, and Robert Healy. 2010. *Knowledge and Environmental Policy: Re-Imagining the Boundaries of Science and Politics*. Cambridge: MIT Press.

Ash, Michael, and T. Robert Fetter. 2004. Who Lives on the Wrong Side of the Environmental Tracks? Evidence from the EPA's Risk-Screening Environmental Indicators Model. *Social Science Quarterly* 85 (2):441–462.

Atlas, Mark. 2007. TRI to Communicate: Public Knowledge of the Federal Toxics Release Inventory. *Social Science Quarterly* 88 (2) (June): 555–572.

Bacot, A. Hunter, and Roy A. Dawes. 1997. State Expenditures and Policy Outcomes in Environmental Program Management. *Policy Studies Journal: the Journal of the Policy Studies Organization* 25 (3):355–370.

Bae, Hyunhoe, Peter Wilcoxen, and David Popp. 2010. Information Disclosure Policy: Do State Data Processing Efforts Help More Than the Information Disclosure Itself? *Journal of Policy Analysis and Management* 29 (1):163–182.

Bardach, Eugene, and Robert A. Kagan. 1982. *Going by the Book: The Problem of Regulatory Unreasonableness*. Philadelphia: Temple University Press.

Barrilleaux, Charles J., and Mark E. Miller. 1988. The Political Economy of State Medicaid Policy. *American Political Science Review* 82 (4):1089–1107.

Baruch, Yehuda, and Brooks C. Holton. 2008. Survey Response Rate Levels and Trends in Organizational Research. *Human Relations* 61 (8):1139–1160.

Baue, Bill. 2008. Investing for Sustainability. In *State of the World 2008: Innovations for a Sustainable Economy*, ed. Linda Starke. New York: W. W. Norton, 180–195.

Beierle, Thomas C. 2003. *The Benefits and Costs of Environmental Information Disclosure: What Do We Know About Right-to-Know?* Washington, D.C.: Resources for the Future, RFF Discussion Paper 03–05.

Beierle, Thomas C., and Jerry Cayford. 2002. *Democracy in Practice: Public Participation in Environmental Decision Making*. Washington, D.C.: RFF Press.

Bennear, Lori Synder. 2007. Are Management-Based Regulations Effective? Evidence from State Pollution Prevention Programs. *Journal of Policy Analysis and Management* 26 (2):327–348.

Berry, Jeffry M. 1997. *The Interest Group Society*. 3rd ed. New York: Longman.

Birkland, Thomas A. 1997. *After Disaster: Agenda Setting, Public Policy, and Focusing Events*. Washington, D.C.: Georgetown University Press.

Borck, Jonathan C., and Cary Coglianese. 2009. Voluntary Environmental Programs: Assessing Their Effectiveness. *Annual Review of Environment and Resources* 34:305–325.

Bosso, Christopher J. 2005. *Environment, Inc.: From Grassroots to Beltway*. Lawrence, KS: University Press of Kansas.

Bosso, Christopher J., and Deborah Lynn Guber. 2006. Maintaining Presence: Environmental Advocacy and the Permanent Campaign. In *Environmental Policy*. 6th ed., ed. Norman J. Vig and Michael E. Kraft. Washington, D.C.: CQ Press, 78–99.

Bouwes, Nicholaas, Steven M. Hassur, and Marc D. Shapiro. 2001. *Empowerment Through Risk-Related Information: EPA's Risk Screening Environmental Indicators Project*. University of Massachusetts, Amherst, Political Economy Research Institute, Working Paper Series Number 18.

Broder, John M. 2009. "Greenhouse Gases Imperil Health, E.P.A. Announces: Ruling Paves Way for Emissions Regulation." *New York Times*, December 8, 2009, online edition.

Carmines, Edward G., and Richard A. Zeller. 1979. *Reliability and Validity Assessment*. Beverly Hills: Sage.

Chemical Week. 1985. We Think We Can Handle This. *Chemical Week*, January 2, 1985, 6.

Chertow, Marian R., and Daniel C. Esty, eds. 1997. *Thinking Ecologically: The Next Generation of Environmental Policy*. New Haven, CT: Yale University Press.

Clark, George E. 2009. Environmental Twitter. *Environment* 51 (5):5–6.

Coase, Ronald. 1960. The Problem of Social Cost. *Journal of Law & Economics* 3 (October):1–44.

Coase, Ronald. 1988. *The Firm, the Market, and the Law*. Chicago: University of Chicago Press.

Coglianese, Cary. 2010. Engaging Business in the Regulation of Nanotechnology. In *Governing Uncertainty: Environmental Regulation in the Age of Nanotechnology*, ed. Christopher J. Bosso. Washington, D.C.: RFF Press, 46–79.

Coglianese, Cary, and Jennifer Nash, eds. 2001. *Regulating from the Inside: Can Environmental Management Systems Achieve Policy Goals*. Washington, D.C.: Resources for the Future Press.

Coglianese, Cary, and Jennifer Nash, eds. 2006a. *Leveraging the Private Sector: Management-Based Strategies for Improving Environmental Performance*. Washington, D.C.: RFF Press.

Coglianese, Cary, and Jennifer Nash. 2006b. *Beyond Compliance: Business Decision Making and the US EPA's Performance Track Program*. Cambridge, MA: Regulatory Policy Program, John F. Kennedy School of Government, Harvard University.

Cohen, Steven. 1984. Defusing the Toxic Time Bomb: Federal Hazardous Waste Programs. In *Environmental Policy in the 1980s*, ed. Norman J. Vig and Michael E. Kraft. Washington, DC: CQ Press, 273–291.

Covello, Vincent T, and Jeryl M. Mumpower. 1986. Risk Analysis and Risk Management: A Historical Perspective. In *Risk Evaluation and Management: Contemporary Issues in Risk Analysis*, ed. Vincent T. Covello, Joshua Menkes, and Jeryl M. Mumpower. New York: Plenum Press, 33–54.

Cowan, Alison Leigh. 2009. "8,000 Federal Forms, 10 Billion Hours, and One Paperwork Reduction Act." *New York Times*, July 13, 2009.

Crabb, Charlene. 2004. Revisiting the Bhopal Tragedy. *Science* 306 (December 3):1670–1671.

Dahl, Richard. 1997. Now That You Know. *Environmental Health Perspectives* 105 (1):38–43.

Daley, Dorothy M. 2007. Citizen Groups and Scientific Decisionmaking: Does Public Participation Influence Environmental Outcomes. *Journal of Policy Analysis and Management* 26 (2):349–368.

Daley, Dorothy M., and James C. Garand. 2005. Horizontal Diffusion, Vertical Diffusion, and Internal Pressure in State Environmental Policy Making, 1989–1998. *American Politics Research* (September):615–644.

Davies, J. Clarence, ed. 1996. *Comparing Environmental Risks: Tools for Setting Government Priorities*. Washington, D.C.: Resources for the Future.

Davies, J. Clarence, and Jan Mazurek. 1998. *Pollution Control in the United States: Evaluating the System*. Washington, D.C.: Resources for the Future.

Davis, Charles E., and Richard Feiock. 1992. Testing Theories of State Hazardous Waste Regulation. *American Politics Quarterly* 20 (4):501–511.

Davis, Charles E., and James P. Lester. 1987. Decentralizing Federal Environmental Policy: A Research Note. *Western Political Quarterly* 40 (3):555–565.

de Marchi, Scott, and James T. Hamilton. 2006. Assessing the Accuracy of Self-Reported Data: An Evaluation of the Toxics Release Inventory. *Journal of Risk and Uncertainty* 32 (1):57–76.

deLeon, Peter, and Jorge E. Rivera eds. 2007. Voluntary Environmental Programs: A Symposium. *Policy Studies Journal: the Journal of the Policy Studies Organization* 35 (4):685–688.

Delmas, Magali A., and Michael E. Toffel. 2004. "Institutional Pressure and Environmental Management Practices: An Empirical Analysis." Paper presented at an EPA-sponsored workshop on Corporate Environmental Behavior and the Effectiveness of Government Interventions, Washington, D.C.: April.

DePalma, Anthony. 2007. "E.P.A. Is Sued by 12 States over Reports on Chemicals." *New York Times*, November 29, 2007, A16.

Deutsch, Claudia H. 2007. "Climate-Change Scorecard Aims to Influence Consumers." *New York Times*, June 19, 2007, C3.

Dietz, Thomas. 2003. Understanding Voluntary Measures. In *New Tools for Environmental Protection: Education, Information, and Voluntary Measures*, ed. Thomas Dietz and Paul C. Stern. Washington, D.C.: National Academy Press, 319–333.

Dietz, Thomas, and Paul C. Stern, eds. 2003. *New Tools for Environmental Protection: Education, Information, and Voluntary Measures*. Washington, D.C.: National Academy Press.

Dietz, Thomas, and Paul C. Stern, eds. 2008. *Public Participation in Environmental Assessment and Decision Making*. Washington, D.C.: National Academy Press.

Dillman, Don A. 2000. *Mail and Internet Surveys: The Tailored Design Method*. 2nd ed. New York: John Wiley.

Donahue, John D. 1999. *The Devolution Revolution: Hazardous Crosscurrents*. New York: Century Foundation.

Douglas, Mary, and Aaron Wildavsky. 1982. *Risk and Culture*. Berkeley: University of California Press.

Downs, Anthony. 1957. *An Economic Theory of Democracy*. New York: Harper and Row.

Duhigg, Charles. 2009a. "Debating Just How Much Weed Killer Is Safe in Your Water Glass." *New York Times*, August 23, 2009, 1, 12.

Duhigg, Charles. 2009b. "Clean Water Laws Neglected, at a Cost." *New York Times*, September 13, 2009, 1, 22–24.

Duhigg, Charles. 2009c. "That Tap Water Is Legal but May Be Unhealthy." *New York Times*, December 17, 2009, 1, A30–31.

Durant, Robert F., Daniel J. Fiorino, and Rosemary O'Leary, eds. 2004. *Environmental Governance Reconsidered: Challenges, Choices, and Opportunities*. Cambridge, MA: MIT Press.

Dye, Thomas. 1966. *Politics, Economics, and the Public: Policy Outcomes in the American States*. Chicago, IL: Rand McNally.

Dye, Thomas R., and Virginia Gray. 1980. *The Determinants of Public Policy*. Lexington, MA: Lexington Books.

Eisner, Marc Allen. 2004. Corporate Environmentalism, Regulatory Reform, and Industry Self-Regulation: Toward Genuine Regulatory Reinvention in the United States. *Governance* 17:145–167.

Eisner, Marc Allen. 2007. *Governing the Environment: The Transformation of Environmental Regulation*. Boulder, CO: Lynne Rienner Publishers.

Eisner, Marc Allen, Jeff Worsham, and Evan J. Ringquist. 2006. *Contemporary Regulatory Policy*. 2nd ed. Boulder, CO: Lynne Rienner Publishers.

Environmental Integrity Project. 2004. *Report: Who's Counting?* Released June 22, and available at the organization's Web site: www.environmentalintegrity .org.

Erikson, Robert, Gerald Wright, and John McIver. 1993. *Statehouse Democracy: Public Opinion and Policy in the American States*. New York: Cambridge University Press.

Esty, Daniel C., and Andrew S. Winston. 2006. *Green to Gold: How Smart Companies Use Environmental Strategy to Innovate, Create Value, and Build Competitive Advantage*. New Haven, CT: Yale University Press.

Fiorino, Daniel J. 2001. Environmental Policy as Learning: A New View of an Old Landscape. *Public Administration Review* 61 (May/June): 322–334.

Fiorino, Daniel J. 2004. Flexibility. In *Environmental Governance Reconsidered: Challenges, Choices, and Opportunities*, ed. Robert F. Durant, Daniel J. Fiorino, and Rosemary O'Leary. Cambridge, MA: MIT Press, 393–425.

Fiorino, Daniel J. 2006. *The New Environmental Regulation*. Cambridge, MA: MIT Press.

Fischhoff, Baruch, Sarah Lichtenstein, Paul Slovic, Stephen L. Derby, and Ralph L. Keeney. 1981. *Acceptable Risk*. New York: Cambridge University Press.

Foltz, David H., and Jean H. Peretz. 1997. Evaluating State Hazardous Waste Reduction Policy. *State and Local Government Review* 29 (3):134–146.

Freeman, A. Myrick. 2006. Economics, Incentives, and Environmental Regulation. In *Environmental Policy: New Directions for the Twenty-First Century*, ed. Norman J. Vig and Michael E. Kraft, 193–214. Washington, D.C.: CQ Press.

Fung, Archon, and Dara O'Rourke. 2000. Reinventing Environmental Regulation from the Grassroots Up: Explaining and Expanding the Success of the Toxics Release Inventory. *Environmental Management* 25 (2):115–127.

Fung, Archon, Mary Graham, and David Weil. 2007. *Full Disclosure: The Promise and Perils of Transparency.* Cambridge, UK: Cambridge University Press.

Gamper-Rabindran, Shanti. 2006. Did the EPA's Voluntary Industrial Toxics Program Reduce Emissions? A GIS Analysis of Distributional Impacts and By-Media Analysis of Substitution. *Journal of Environmental Economics and Management* 52:391–410.

Gerber, Brian J., and Paul Teske. 2000. Regulatory Policymaking in the American States: A Review of Theories and Evidence. *Political Research Quarterly* 53 (4):849–886.

Gilbert, Carol Bengle. 2008. "USA Today Reports Toxic Air Surrounds American Schools: Check Your Local School." Available at www.associatedcontent .com.

Gill, Jeff, and Kenneth J. Meier. 2000. Public Administration Research and Practice: A Methodological Manifesto. *Journal of Public Administration: Research and Theory* 10 (1):157–199.

Goetze, David, and C. K. Rowland. 1985. Explaining Hazardous Waste Regulation at the State Level. *Policy Studies Journal: the Journal of the Policy Studies Organization* 14 (1):111–120.

Gormley, William T., Jr. 1989. *Taming the Bureaucracy: Muscles, Prayers, and Other Strategies.* Princeton, N.J.: Princeton University Press.

Gormley, William T., Jr., and Steven J. Balla. 2008. *Bureaucracy and Democracy: Accountability and Performance.* 2nd ed. Washington, D.C.: CQ Press.

Gormley, William T., Jr., and David L. Weimer. 1999. *Organizational Report Cards.* Cambridge: Harvard University Press.

Graham, Mary, and Catherine Miller. 2001. Disclosure of Toxic Releases in the United States. *Environment* 43 (8):8–20.

Graham, Mary. 1999. *The Morning After Earth Day: Practical Environmental Politics.* Washington, D.C.: Brookings Institution Press.

Graham, Mary. 2002. *Democracy by Disclosure.* Washington, D.C.: Brookings Institution Press.

Grant, Don. 1997. Allowing Citizen Participation in Environmental Regulation: An Empirical Analysis of the Effects of Right-to-Sue and Right-to-Know Provisions on Industry's Toxic Emissions. *Social Science Quarterly* 78 (4): 859–873.

Grant, Don, and Andrew W. Jones. 2004. Do Foreign Owned Plants Pollute More? New Evidence from the EPA's Toxics Release Inventory. *Society and Natural Resources* 17 (2): 171–179.

Grant, Don, and Andrew W. Jones. 2003. Are Subsidiaries More Prone to Pollute? New Evidence from the EPA's Toxics Release Inventory. *Social Science Quarterly* 84 (1): 162–173.

Grant, Don, and Andrew W. Jones. 2004. Do Manufacturers Pollute Less Under the Regulation-Through Information Regime? What Plant-Level Data Tell Us. *The Sociological Quarterly* 45 (3): 471–486.

Grant, Don, Andrew W. Jones, and Albert Bergesen. 2002. Organizational Size and Pollution: The Case of the U.S. Chemical Industry. *American Sociological Review* 67: 389–407.

Grant, Don, Andrew W. Jones, and Mary Nell Trautner. 2004. Do Facilities with Distant Headquarters Pollute More? How Civic Engagement Conditions the Environmental Performance of Absentee Plants. *Social Forces* 83 (1): 189–214.

Gray, Virginia. 1973. Innovation in the States: A Diffusion Study. *American Political Science Review* 67 (4):1174–1185.

Guber, Deborah Lynn, and Christopher J. Bosso. 2010. Past the Tipping Point? Public Discourse and the Role of the Environmental Movement in a Post-Bush Era. In *Environmental Policy.* 7th ed., ed. Norman J. Vig and Michael E. Kraft, 51–74. Washington, D.C.: CQ Press.

Guber, Deborah Lynn. 2003. *The Grassroots of a Green Revolution: Polling America on the Environment*. Cambridge: MIT Press.

Gunningham, Neil. 2009. Shaping Corporate Environmental Performance: A Review. *Environmental Policy and Governance* 19 (4):215–231.

Gunningham, Neil, Robert A. Kagen, and Dorothy Thornton. 2003. *Shades of Green: Business, Regulation, and Environment*. Stanford, CA: Stanford University Press.

Hadden, Susan G. 1986. *Read the Label: Reducing Risk by Providing Information*. Boulder, CO: Westview Press.

Hadden, Susan G. 1989. *A Citizen's Right to Know: Risk Communication and Public Policy*. Boulder, CO: Westview Press.

Hadden, Susan G. 1991. Public Perception of Hazardous Waste. *Risk Analysis* 11 (1):47–57.

Hall, Bob, and Mary Lee Kerr. 1991. *1991–1992 Green Index: A State by State Guide to the Nation's Environmental Health*. Washington, D.C.: Island Press.

Hamill, Sean D. 2009a. "Trying to Limit Disclosure on Explosion." *New York Times*, March 29, 2009, 19.

Hamill, Sean D. 2009b. "Plant to Cut Production of Toxic Chemical," *New York Times*, August 27, 2009, A23.

Hamilton, James T. 1995. Pollution as News: Media and Stock Market Reactions to the Toxics Release Inventory Data. *Journal of Environmental Economics and Management* 28:98–113.

Hamilton, James T. 2005. *Regulation Through Revelation: The Origins, Politics, and Impacts of the Toxics Release Inventory Program*. New York: Cambridge University Press.

Hamilton, Lawrence C. 1992. *Regression with Graphics: A Second Course in Applied Statistics*. Belmont, CA: Duxbury Press.

Hanson, Russell L. 1998. The Interaction of State and Local Governments. In *Governing Partners: State-Local Relations in the United States*, ed. Russell L. Hanson. Boulder, CO: Westview Press.

Harrington, Winston, Richard Morgenstern, and Thomas Sterner, eds. 2004. *Choosing Environmental Policy: Comparing Instruments and Outcomes in the United States and Europe*. Washington, D.C.: RFF Press.

Harris, Richard A., and Sidney M. Milkis. 1996. *The Politics of Regulatory Change: A Tale of Two Agencies*. 2nd ed. New York: Oxford University Press.

Harrison, Kathryn. 2003. Challenges in Evaluating Voluntary Environmental Programs. In *New Tools for Environmental Protection: Education, Information, and Voluntary Measures*, ed. Thomas Dietz and Paul C. Stern. Washington, D.C.: National Academy Press, 263–282.

Harrison, Kathryn, and Werner Antweiler. 2003. Incentives for Pollution Abatement: Regulation, Regulatory Threats, and Non-Governmental Pressures. *Journal of Policy Analysis and Management* 22 (3):361–382.

Haskell, Steven. 2009. "Implementing the Emergency Planning and Community Right-to-Know Act through Local Emergency Planning Committees." Master's Thesis, University of Wisconsin Green Bay.

Hays, Scott P., Michael Elser, and Carol F. Hays. 1996. Environmental Commitment among the States: Integrating Alternative Approaches to State Environmental Policy. *Publius* 26 (2):41–58.

Heilprin, John. 2006. "Report Faults EPA for Being Lax on Clean Air Requirements." Associated Press release, July 27, 2006.

Herb, Jeanne, Susan Helms, and Michael J. Jensen. 2003. Harnessing the 'Power of Information': Environmental Right to Know as a Driver of Sound Environmental Policy. In *New Tools for Environmental Protection: Education, Information, and Voluntary Measures*, ed. Thomas Dietz and Paul C. Stern. Washington, D.C.: National Academy Press, 253–262.

Hofferbert, Richard. 1966. The Relationship between Public Policy and Some Structural and Environmental Variables in the American States. *American Political Science Review* 60 (1):73–82.

Hoffman, Andrew J. 2000. *Competitive Environmental Strategy*. New York: Island Press.

Jacobson, Gary C. 2009. *The Politics of Congressional Elections*. 7th ed. New York: Longman.

Jacoby, William G., and Saundra K. Schneider. 2001. Variability in State Policy Priorities: An Empirical Analysis. *Journal of Politics* 63 (2):544–568.

John, DeWitt. 2004. Civic Environmentalism. In *Environmental Governance Reconsidered: Challenges, Choices, and Opportunities*, ed. Robert F. Durant, Daniel J. Fiorino, and Rosemary O'Leary, 219–254. Cambridge, MA: MIT Press.

Johnson, Janet Buttolph, and H. T. Reynolds. 2005. *Political Science Research Methods*. 5th ed. Washington, D.C.: CQ Press.

Kaufman, Leslie. 2009a. "Big Polluters Told to Report Emissions." *New York Times*, September 23, 2009, B2.

Kaufman, Leslie. 2009b. "Coming Clean About Carbon: Industries Disclose Emissions to Claim the High Ground." *New York Times*, December 29, 2009, B1–8.

Keller, Ann Campbell. 2009. *Science in Environmental Policy: The Politics of Objective Advice*. Cambridge: MIT Press.

Kettl, Donald F., ed. 2002. *Environmental Governance: A Report on the Next Generation of Environmental Policy*. Washington, D.C.: Brookings Institution Press.

Khanna, Madhu, Rose H. Quimio, and Dora Bojilova. 1998. Toxic Release Information: A Policy Tool for Environmental Protection. *Journal of Environmental Economics and Management* 36:243–266.

King, Gary. 1986. How Not to Lie with Statistics: Avoiding Common Mistakes in Quantitative Political Science. *American Journal of Political Science* 30 (3):666–687.

King, Gary, Michael Tomz, and Jason Wittenberg. 2000. Making the Most of Statistical Analysis: Improving Interpretation and Presentation. *American Journal of Political Science* 44 (2):347–361.

Kingdon, John W. 1995. *Agendas, Alternatives, and Public Policies*. 2nd ed. New York: Longman.

Klyza, Christopher McGrory, and David Sousa. 2008. *American Environmental Policy, 1990–2006: Beyond Gridlock*. Cambridge, MA: MIT Press.

Konar, Shameek, and Mark A. Cohen. 1997. Information as Regulation: The Effect of Community Right to Know Laws on Toxic Emissions. *Journal of Environmental Economics and Management* 32 (1):109–124.

Konisky, David. 2007. Regulatory Competition and Environmental Enforcement: Is There a Race to the Bottom? *American Journal of Political Science* 51 (4):853–872.

Koontz, Tomas M. 2002. State Innovation in Natural Resources Policy: Ecosystem Management on Public Forests. *State and Local Government Review* 34 (3):160–172.

Kraft, Michael E. 1998. Using Environmental Program Evaluation: Politics, Knowledge, and Policy Change. In *Environmental Program Evaluation: A Primer*, ed. Gerrit J. Knaap and T. John Kim. Urbana: University of Illinois Press, 293–320.

Kraft, Michael E. 2010. Environmental Policy in Congress. In *Environmental Policy: New Directions for the Twenty-First Century*. 7th ed., ed. Norman J. Vig and Michael E. Kraft. Washington, D.C.: CQ Press, 99–124.

Kraft, Michael E. 2011. *Environmental Policy and Politics*. 5th ed. New York: Pearson Longman.

Kraft, Michael E., and Scott R. Furlong. 2010. *Public Policy: Politics, Analysis, and Alternatives*. 3rd ed. Washington, D.C.: CQ Press.

Kraft, Michael E., and Sheldon Kamieniecki, eds. 2007. *Business and Environmental Policy: Corporate Interests in the American Political System*. Cambridge, MA: MIT Press.

Kriz, Margaret. 1988. Fuming Over Fumes. *National Journal* (November 26), 3006–3009.

Laszlo, Chris. 2003. *The Sustainable Company: How to Create Lasting Value Through Social and Environmental Performance*. Washington, D.C.: Island Press.

Layton, Lyndsey. 2010. "Use of Potentially Harmful Chemicals Kept Secret Under Law." *Washington Post*, January 4, 2010, A01.

Layzer, Judith. 2008. *Natural Experiments: Ecosystem Management and the Environment*. Cambridge, MA: MIT Press.

Leonnig, Carol D., Jo Becker, and David Nakamura. 2004. "Misleading Results: Utilities Are Manipulating Data on Lead Levels in Water to Skirt Federal Regulations." *Washington Post National Weekly Edition*, October 18–24, 2004, 6–7.

Lester, James P. 1994. A New Federalism? Environmental Policy in the States. In *Environmental Policy in the 1990s*. 2nd ed., ed. Norman J. Vig and Michael E. Kraft. Washington: CQ Press, 51–68.

Lester, James P. 1986. New Federalism and Environmental Policy. *Publius* 16 (1):149–165.

Lester, James P., and Ann O'M. Bowman, eds. 1983. *The Politics of Hazardous Waste Management*. Durham, NC: Duke University Press.

Lester, James P., David W. Allen, and Kelly M. Hill. 2001. *Environmental Injustice in the United States: Myths and Realities*. Boulder: Westview Press.

Lester, James P., and James L. Franke, Ann O'M. Bowman, and Kenneth W. Kramer. 1983a. A Comparative Perspective on State Hazardous Waste Regulation. In *The Politics of Hazardous Waste Management*, ed. James P. Lester and Ann O'M. Bowman, 212–231. Durham: Duke University Press.

Lester, James P., James L. Franke, Ann O'M Bowman, and Kenneth W. Kramer. 1983b. Hazardous Waste, Politics, and Policy: A Comparative State Analysis. *Western Political Science Quarterly* 36:257–285.

Lewis-Beck, Michael S. 1977. The Relative Importance of Socioeconomic and Political Variables for Public Policy. *American Political Science Review* 71 (2):559–566.

Lewis-Beck, Michael S. 1980. *Applied Regression: An Introduction*. Beverly Hills, London: Sage Publications.

Lichtblau, Eric. 2008. "In Justice Shift, Corporate Deals Replace Trials." *New York Times*, April 9, 2008, 1, A20.

Lindblom, Charles E. 1980. *The Policy-Making Process*. 2nd ed. Englewood Cliffs, NJ: Prentice Hall.

Lindblom, Charles E., and David K. Cohen. 1979. *Usable Knowledge: Social Science and Social Problem Solving*. New Haven, CT: Yale University Press.

Lipton, Eric. 2005. "Administration to Seek Antiterror Rules for Chemical Plants." *New York Times,* June 15, 2005, A18.

Lipton, Eric. 2006a. "Chertoff Seeks a Chemical Security Law, Within Limits." *New York Times,* March 22, 2006, A18.

Lipton, Eric. 2006b. "Citing Security, Plants Use Safer Chemicals." *New York Times,* April 25, 2006, A20.

Lombard, Emmett N. 1993. Intergovernmental Relations and Air Quality Policy. *American Review of Public Administration* 23 (1):57–73.

Lowrance, William W. 1976. *Of Acceptable Risk: Science and the Determination of Safety.* Los Altos, CA: William Kaufman.

Lowry, William. 1992. *The Dimensions of Federalism: State Governments and Pollution Control Policies.* Durham: Duke University Press.

Lubell, Mark. 2004. Collaborative Environmental Institutions: All Talk and No Action? *Journal of Policy Analysis and Management* 23 (3):549–573.

Lynch, Barbara D. 1989. *Right-to-Know: An Opportunity to Learn.* Proceedings of the Institute for Comparative and Environmental Toxicology Symposium IV. Ithaca, New York: Cornell University, October 5–6.

Lynn, Frances M., and Jack D. Kartez. 1994. Environmental Democracy in Action: The Toxics Release Inventory. *Environmental Management* 18 (4):511–521.

Lynn, Frances M., George Busenberg, Nevin Cohen, and Caron Chess. 2000. Chemical Industry's Community Advisory Panels: What Has Been Their Impact? *Environmental Science & Technology* 34 (10):1881–1886.

Lynn, Laurence E., ed. 1978. *Knowledge and Policy: The Uncertain Connection.* Washington, D.C.: National Academy of Sciences.

Lyon, Thomas P., and John W. Maxwell. 2004. *Corporate Environmentalism and Public Policy.* Cambridge, UK: Cambridge University Press.

MacLean, Alair, and Paul Orum. 1992. *Progress Report: Community Right-to-Know.* Working Group on Community Right-to-Know, Washington, D.C., July.

Marcus, Alfred A., Donald A. Geffen, and Ken Sexton. 2002. *Reinventing Environmental Regulation: Lessons from Project XL.* Washington, D.C.: RFF Press.

May, Peter, and Chris Koski. 2007. State Environmental Policies: Analyzing Green Building Mandates. *Review of Policy Research* 24 (1):49–65.

Mayer, Brian, Phil Brown, and Meadow Linder. 2002. Moving Further Upstream: From Toxics Reduction to the Precautionary Principle. *Public Health Reports* 117 (6):574–586.

Matheny, Albert R., and Bruce Williams. 1981. Scientific Disputes and Adversary Procedures in Policy-Making: An Evaluation of the Science Court. *Law & Policy Quarterly* 3 (3):341–364.

Mayhew, David R. 1974. *Congress: The Electoral Connection.* New Haven, CT: Yale University Press.

Mazmanian, Daniel A., and Michael E. Kraft, eds. 2009. *Toward Sustainable Communities: Transition and Transformations in Environmental Policy.* 2nd ed. Cambridge, MA: MIT Press.

Mazurek, Janice. 2003. Government-Sponsored Voluntary Programs for Firms: An Initial Survey. In *New Tools for Environmental Protection: Education, Information, and Voluntary Measures,* ed. Thomas Dietz and Paul C. Stern. Washington, D.C.: National Academy Press, 219–233.

Meier, Kenneth J., and Jeff Gill. 2000. *What Works? A New Approach to Program and Policy Analysis.* Boulder, CO: Westview Press.

Meier, Kenneth J., and Lael R. Keiser. 1996. Public Administration as a Science of the Artificial: A Methodology for Prescription. *Public Administration Review* 56 (September/October):459–466.

Metzenbaum, Shelley H. 2001. Information, Environmental Performance, and Environmental Management Systems. In *Regulating from the Inside,* ed. Cary Coglianese and Jennifer Nash. Washington, D.C.: RFF Press, 146–180.

Meyers, Robert J. 2008. "EPA Makes Steady Progress: Agency Has a Variety of Programs to Protect Children at Their Schools." *USA Today,* December 11, 2008.

Morgenstern, Richard D., and William A. Pizer, eds. 2007. *Reality Check: The Nature and Performance of Voluntary Environmental Programs in the United States, Europe, and Japan.* Washington, D.C.: RFF Press.

Morrison, Blake, and Brad Heath. 2008a. "Toxic Air and America's Schools: Health Risks Stack up for School Kids Near Industry: Analysis Pinpoints Toxic Hot Spots in 34 States." *USA Today,* December 8, 2008.

Morrison, Blake, and Brad Heath. 2008b. "Officials Vow Air Will Be Tested: Senator, States Pledge Action Near Schools." *USA Today,* December 10, 2008.

Natan, Thomas E., Jr., and Catherine G. Miller. 1998. Are Toxics Release Inventory Reductions Real? *Environmental Science & Technology* 32 (15):368A–374A.

National Academy of Public Administration (NAPA). 1995. *Setting Priorities, Getting Results: A New Direction for EPA.* Washington, D.C.: National Academy of Public Administration.

National Academy of Public Administration (NAPA). 2000. *Environment.gov: Transforming Environmental Protection for the 21st Century.* Washington, D.C.: National Academy of Public Administration.

National Research Council (Committee on Risk Perception and Communication). 1989. *Improving Risk Communication.* Washington, D.C.: National Academy Press.

Nice, David C. 1998. The Intergovernmental Setting of State-Local Relations. In *Governing Partners: State-Local Relations in the United States,* ed. Russell L. Hanson. Boulder, CO: Westview Press, 17–36.

Nice, David, and Patricia Frederickson. 1995. *The Politics of Intergovernmental Relations.* 2nd ed. Chicago: Nelson-Hall.

O'Rourke, Dara, and Eungkyoon Lee. 2004. Mandatory Planning for Environmental Innovation: Evaluating Regulatory Mechanisms for Toxic Use Reduction. *Journal of Environmental Planning and Management* 47 (2):181–200.

O'Rourke, Dara, and Gregg P. Macey. 2003. Community Environmental Policing: Assessing New Strategies of Public Participation in Environmental Regulation. *Journal of Policy Analysis and Management* 22 (3):383–414.

O'Toole, Laurence J., Chilik Yu, James Cooley, Gail Cowie, Susan Crow, Terry DeMeo, and Stephanie Herbert. 1997. Reducing Toxic Chemical Releases and Transfers: Explaining Outcomes for a Voluntary Program. *Policy Studies Journal: the Journal of the Policy Studies Organization* 25 (1):11–26.

Olmstead, Sheila M. 2010. Applying Market Principles to Environmental Policy. In *Environmental Policy: New Directions for the Twenty-First Century*, ed. Norman J. Vig and Michael E. Kraft. Washington, D.C.: CQ Press, 197–219.

Olson, Mancur. 1971. *The Logic of Collective Action: Public Goods and the Theory of Groups*. Cambridge, MA: Harvard University Press.

Pastor, Manual, Jr., James L. Sadd, and Rachel Morello-Frosch. 2004. Waiting to Inhale: The Demographics of Toxic Air Release Facilities in 21st Century California. *Social Science Quarterly* 85 (2):420–440.

Perrow, Charles. 1999. *Normal Accidents: Living with High-Risk Technologies, updated edition*. Princeton, NJ: Princeton University Press.

Portney, Paul R. 1998. Counting the Cost: The Growing Role of Economics in Environmental Decisionmaking. *Environment* 40 (2):12–18, 36–38.

Portney, Paul R., and Katherine N. Probst. 1994. Cleaning Up Superfund. *Resources* 114 (Winter):2–5.

Portney, Paul R., and Robert N. Stavins, eds. 2000. *Public Policies for Environmental Protection*. 2nd ed. Washington, D.C.: Resources for the Future Press.

Potoski, Matthew. 2001. Clean Air Federalism: Do States Race to the Bottom? *Public Administration Review* 61 (3):335–343.

Potoski, Matthew, and Aseem Prakash. 2004. The Regulation Dilemma: Cooperation and Conflict in Environmental Governance. *Public Administration Review* 64 (2):152–163.

Potoski, Matthew, and Aseem Prakash. 2005. Covenants with Weak Swords: ISO 14001 and Facilities' Environmental Performance. *Journal of Policy Analysis and Management* 24 (4):745–769.

Potoski, Matthew, and Aseem Prakash, eds. 2009. *Voluntary Programs: A Club Theory Perspective*. Cambridge, MA: MIT Press.

Potoski, Matthew, and Neal D. Woods. 2002. Dimensions of State Environmental Policies: Air Pollution Regulation in the United States. *Policy Studies Journal: the Journal of the Policy Studies Organization* 30 (2):208–226.

Powell, Mark R. 1999. *Science at EPA: Information in the Regulatory Process*. Washington, D.C.: Resources for the Future Press.

Prakash, Aseem. 2000. *Greening the Firm: The Politics of Corporate Environmentalism*. Cambridge, U.K.: Cambridge University Press.

Prakash, Aseem. 2003. Factors in Firms and Industries Affecting the Outcomes of Voluntary Measures. In *New Tools for Environmental Protection: Education, Information, and Voluntary Measures*, ed. Thomas Dietz and Paul C. Stern. Washington, D.C.: National Academy Press, 303–310.

Prakash, Aseem, and Matthew Potoski. 2006. *The Voluntary Environmentalists: Green Clubs, ISO 14001, and Voluntary Environmental Regulations*. New York: Cambridge University Press.

Press, Daniel. 1998. Local Environmental Policy Capacity: A Framework for Research. *Natural Resources Journal* 38 (1):29–52.

Press, Daniel. 1999. Local Open Space Preservation in California. In *Toward Sustainable Communities*, ed. Daniel A. Mazmanian and Michael E. Kraft. Cambridge: MIT Press, 153–183.

Press, Daniel. 2007. Industry, Environmental Policy, and Environmental Outcomes. *Annual Review of Environment and Resources* 32:1.1–1.28.

Press, Daniel, and Daniel A. Mazmanian. 1997. The Greening of Industry: Achievement and Potential. In *Environmental Policy in the 1990s*. 3rd ed., ed. Norman J. Vig and Michael E. Kraft. Washington, D.C.: CQ Press, 255–277.

Press, Daniel, and Daniel A. Mazmanian. 2010. Toward Sustainable Production: Finding Workable Strategies for Government and Industry. In *Environmental Policy*. 7th ed., ed. Norman J. Vig and Michael E. Kraft. Washington, D.C.: CQ Press, 220–243.

Rabe, Barry G. 1991. Environmental Regulation in New Jersey: Innovations and Limitations. *Publius* 21 (1):83–103.

Rabe, Barry G. 1995. Integrating Environmental Regulation: Permitting Innovation at the State Level. *Journal of Policy Analysis and Management* 14 (3):467–472.

Rabe, Barry G. 1996. An Empirical Examination of Innovations in Integrated Environmental Management: The Case of the Great Lakes Basin. *Public Administration Review* 56 (4):373–381.

Rabe, Barry G. 1999. Federalism and Entrepreneurship: Explaining American and Canadian Innovation in Pollution Prevention and Regulatory Integration. *Policy Studies Journal: the Journal of the Policy Studies Organization* 27 (2):288–306.

Rabe, Barry G. 2010. Racing to the Top, the Bottom or the Middle of the Pack? The Evolving State Government Role in Environmental Protection? In *Environmental Policy: New Directions for the Twenty-First Century*. 7th ed., ed. Norman J. Vig and Michael E. Kraft. Washington, D.C.: CQ Press, 27–50.

Rabe, Barry G., and Marc Gaden. 2009. Sustainability in a Regional Context: The Case of the Great Lakes Basin. In *Toward Sustainable Communities*. 2nd ed., ed. Daniel A. Mazmanian and Michael E. Kraft. Cambridge: MIT Press, 289–314.

Rapoport, Anatol, and Albert M. Chammah. 1965. *Prisoner's Dilemma*. Ann Arbor: The University of Michigan Press.

Reibstein, Rick. 2008. "What If Technical Assistance Really Works?" *Sustain*, Fall/Winter.

Richardson, John. 2006. "Report: State's Toxic Releases Up; Government Findings Show a 13 Percent Jump in Maine, but Industry Officials Say It's Due to Better Tracking." *Portland Press Herald*, April 30, 2006, A1.

Ringquist, Evan J. 1993a. *Environmental Protection at the State Level: Politics and Progress in Controlling Pollution*. Armonk, New York: Sharpe.

Ringquist, Evan J. 1993b. Does Regulation Matter? Evaluating the Effects of State Air Pollution Control Programs. *Journal of Politics* 55 (4):1022–1045.

Ringquist, Evan J. 1994. Policy Influence and Policy Responsiveness in State Pollution Control. *Policy Studies Journal: the Journal of the Policy Studies Organization* 22 (1):25–43.

Ringquist, Evan J. 1995. Is 'Effective Regulation' Always Oxymoronic? The States and Ambient Air Quality. *Social Science Quarterly* 76 (1):69–87.

Ringquist, Evan J., and David H. Clark. 2002. Issue Definition and the Politics of State Environmental Justice Policy Adoption. *International Journal of Public Administration* 25 (2, 3):351–389.

Ryer, Eric. 2007. "Demonstrating the Empirical Relationship Between Corporate Social Performance, Firm Size, and Environmental and Financial Performance." Oshkosh, WI: paper prepared for the University of Wisconsin-Oshkosh Master of Public Administration Program.

Sabatier, Paul. 1978. The Acquisition and Utilization of Technical Information by Administrative Agencies. *Administrative Science Quarterly* 23 (Sept): 386–411.

Sabatier, Paul A., ed. 2007. *Theories of the Policy Process*. 2nd ed. Boulder, CO: Westview Press.

Sabatier, Paul A., and Hank C. Jenkins-Smith. 1993. *Policy Change and Learning: An Advocacy Coalition Approach*. Boulder, CO: Westview Press.

Sabatier, Paul A., Will Focht, Mark Lubell, Zev Trachtenberg, Arnold Vedlitz, and Marty Matlock, eds. 2005. *Swimming Upstream: Collaborative Approaches to Watershed Management*. Cambridge, MA: MIT Press.

Santos, Susan L., Vincent T. Covello, and David B. McCallum. 1996. Industry Response to SARA Title III: Pollution Prevention, Risk Reduction, and Risk Communication. *Risk Analysis* 16 (1):57–66.

Sapat, Alka. 2004. Devolution and Innovation: The Adoption of State Environmental Policy Innovations by Administrative Agencies. *Public Administration Review* 64 (2):141–151.

Sarokin, David, and Jay Schulkin. 1991. Environmentalism and the Right-to-Know: Expanding the Practice of Democracy. *Ecological Economics* 4:175–189.

Scheberle, Denise. 2004. *Federalism and Environmental Policy: Public Trust and the Politics of Implementation*. 2nd ed. Washington, D.C.: Georgetown University Press.

Schmidt, Charles W. 2003. The Risk Where You Live. *Environmental Health Perspectives* 111 (7):A404–A407.

Schneider, Anne Larason, and Helen Ingram. 1997. *Policy Design for Democracy*. Lawrence: University of Kansas Press.

Schoenbrod, David. 2005. *Saving Our Environment from Washington: How Congress Grabs Power, Shirks Responsibility, and Shortchanges the People*. New Haven, CT: Yale University Press.

Schoenbrod, David, Richard Stewart, and Katrina Wyman. 2009. "Breaking the Logjam: Environmental Reform for the New Congress and Administration." Manuscript, available at www.breakingthelogjam.org.

Scholz, John T. 1984. Voluntary Compliance and Regulatory Enforcement. *Law & Policy* 6:385–404.

Scholz, John T. 1991. Cooperative Regulatory Enforcement and the Politics of Administrative Effectiveness. *American Political Science Review* 85 (1): 115–136.

Scholz, John T., and Wayne Gray. 1997. Can Government Facilitate Cooperation? An Informational Model of OSHA Enforcement. *American Journal of Political Science* 41 (3):693–717.

Schwarzman, Megan R., and Michael P. Wilson. 2009. New Science for Chemicals Policy. *Science* 326 (November 20): 1065–1066.

Sengupta, Somini. 2008. "Decades Later, Toxic Sludge Torments Bhopal," *New York Times*, July 7, 2008, A1, A8.

Sexton, Ken, Alfred A. Marcus, K. William Easter, and Timothy D. Burkhardt, eds. 1999. *Better Environmental Decisions: Strategies for Governments, Businesses, and Communities*. Washington, D.C.: Island Press.

Shapiro, Marc D. 2005. Equity and Information: Information Regulation, Environmental Justice, and Risks from Toxic Chemicals. *Journal of Policy Analysis and Management* 24 (2):373–398.

Shapiro, Michael. 1990. Toxic Substances Policy. In *Public Policies for Environmental Protection*, ed. Paul R. Portney. Washington, D.C.: Resources for the Future Press, 195–241.

Shebecoff, Philip. 1989. "U.S. Only Narrowly Avoided 17 Bhopal-Like Disasters, Study Says." *New York Times*, April 30, 1989, 16.

Shneiderman, Ben. 2009. Civic Collaboration. *Science* 325 (July 31): 540.

Shrader-Frechette, K. S. 1991. *Risk and Rationality: Philosophical Foundations for Populist Reforms*. Berkeley: University of California Press.

Sheiman, Deborah. 1991. *The Right to Know More*. Washington, D.C.: Natural Resources Defense Council, May.

Sigman, Hillary. 2000. Hazardous Waste and Toxic Substance Policies. In *Public Policies for Environmental Protection*. 2nd ed., ed. Paul R. Portney and Robert N. Stavins. Washington, D.C.: RFF Press, 215–259.

Sigman, Hilary. 2003. *Letting States Do the Dirty Work: State Responsibility for Federal Environmental Regulation*. Unpublished manuscript, National Bureau of Economic Research.

Skrzycki, Cindy. 2006. "Chemical-Data Plan Catalyzes Opposition," *Washington Post*, January 3, 2006, D1.

Slovic, Paul. 1987. Perceptions of Risk. *Science* 236:280–285.

Slovic, Paul. 1993. Perceived Risk, Trust, and Democracy. *Risk Analysis* 13:675–682.

Smith, Eric R. A. N. 2002. *Energy, the Environment, and Public Opinion*. Boulder, CO: Rowman and Littlefield.

Speth, James Gustave. 2008. *The Bridge at the Edge of the World: Capitalism, the Environment, and Crossing from Crisis to Sustainability*. New Haven: Yale University Press.

Stephan, Mark. 2002. Environmental Information Disclosure Programs: They Work, but Why? *Social Science Quarterly* 83:190–205.

Stephan, Mark, Michael E. Kraft, and Troy D. Abel. 2005. "Information Politics and Environmental Performance: The Impact of the Toxics Release Inventory on Corporate Decision Making." Paper presented at the annual meeting of the American Political Science Association. September 1–4, Washington, D.C.

Stern, Paul C., and Harvey V. Fineberg, eds. 1996. *Understanding Risk: Informing Decisions in a Democratic Society*. Washington, D.C.: National Academy Press. Committee on Risk Characterization, Commission on Behavioral and Social Sciences and Education, National Research Council.

Teske, Paul. 2004. *Regulation in the States*. Washington, D.C.: Brookings Institution Press.

Tietenberg, Tom, and David Wheeler. 1998. "Empowering the Community: Information Strategies for Pollution Control." In *Frontiers of Environmental Economics Conference*. (www.worldbank.org/nipr/work_paper/ecoenv/index.htm).

U.S. Bureau of the Census. 1993. *The Statistical Abstract of the United States*.

U.S. Environmental Protection Agency. 1990. *Reducing Risk: Setting Priorities and Strategies for Environmental Protection*. Washington, D.C.: EPA, Science Advisory Board.

U.S. Environmental Protection Agency. 1993. *1991 Toxics Release Inventory (TRI) Public Data Release*. Office of Pollution Prevention and Toxics.

U.S. Environmental Protection Agency. 1997. *The Benefits and Costs of the Clean Air Act, 1970 to 1990*. Washington, D.C.: Office of Policy, Planning, and Evaluation and Office of Air and Radiation, EPA. Available at www.epa.gov/air/sect812/copy.html.

U.S. Environmental Protection Agency. 1998. *1996 Toxics Release Inventory (TRI) Public Data Release*. Office of Pollution Prevention and Toxics.

U.S. Environmental Protection Agency. 1999. *Enforcement and Compliance Assurance FY98 Accomplishments Report*. Office of Enforcement and Compliance Assurance. Washington, D.C.: U.S. EPA, June.

U.S. Environmental Protection Agency. 2002. Factors to Consider When Using TRI Data, EPA-260-F-02-017.

U.S. Environmental Protection Agency. 2003. "How Are the Toxics Release Inventory Data Used? Government, Business, Academic and Citizen Uses." Office of Environmental Information, Office of Information Analysis and Access, May. Available at www.epa.gov/tri/guide_docs/pdf/2003/2003_datausepaper.pdf.

U.S. Environmental Protection Agency. 2004. *Community Air Screening How-To Manual, A Step-by-Step Guide to Using Risk-Based Screening to Identify Priorities for Improving Outdoor Air Quality.* Washington, D.C.: EPA Office of Pollution Prevention and Toxic Substances.

U.S. Environmental Protection Agency. 2007. *2005 Toxics Release Inventory (TRI) Public Data Release Report.* Washington, D.C.: EPA Office of Environmental Information, May.

U.S. Environmental Protection Agency. 2008. "National Air Quality—Status and Trends through 2007" (November), available at http://www.epa.gov/air/airtrends/2008/.

U.S. Environmental Protection Agency. 2009. *2007 TRI Public Data Release.* Washington, D.C.: U.S. EPA., March 19, available at: http://www.epa.gov/tri/tridata/tri07/index.htm.

U.S. General Accounting Office (now Government Accountability Office) 1991. *Toxic Chemicals: EPA's Toxic Release Inventory Is Useful But Can Be Improved.* Washington, D.C.: U.S. GAO, GAO/RECD-91-121.

U.S. Government Accountability Office. 2006. *EPA Should Improve the Management of Its Air Toxics Program.* Washington, D.C.: U.S. GAO, GAO-06-669.

U.S. Government Accountability Office. 2007a. "Environmental Right-to-Know: EPA's Recent Rule Could Reduce Availability of Toxic Chemical Information Used to Assess Environmental Justice." Washington, D.C.: GAO, GAO-08-115T, July.

U.S. Government Accountability Office. 2007b. *EPA's Recent Rule Could Reduce the Availability of Toxic Chemical Information Used to Assess Environmental Justice.* Washington, D.C.: GAO, GAO-08-115T, October 4.

U.S. Senate. 1985. Committee on Small Business. *Community Right to Know Legislation and Its Regulatory and Paperwork Impact on Small Business.* 99th Congress, 1st sess. 18 June.

Van Horn, Carl E. 1996. The Quiet Revolution. In *The State of the States.* 3rd ed., ed. Carl E. Van Horn. Washington, D.C.: Congressional Quarterly.

Vig, Norman J., and Michael E. Kraft, eds. 1984. *Environmental Policy in the 1980s: Reagan's New Agenda.* Washington, D.C.: CQ Press.

Vig, Norman J., and Michael E. Kraft, eds. 2010. *Environmental Policy.* 7th ed. Washington, D.C.: CQ Press.

Vogel, David. 2005. *The Market for Virtue: The Potential and Limits of Corporate Social Responsibility.* Washington, D.C.: Brookings Institution Press.

Wald, Matthew L. 2009a. "A Drop in Toxic Emissions and a Rise in PCB Disposal." *New York Times,* March 20, 2009, A15.

Wald, Matthew L. 2009b. "Lawmakers Say Chemical Company Withheld Information About Explosion." *New York Times,* April 22, 2009, A13.

Walker, Jack L. 1969. The Diffusion of Innovations among the American States. *American Political Science Review* 63 (3):880–899.

Wang, Hua, Jun Bi, David Wheeler, Jinnan Wang, Dong Cao, Genfa Lu, and Yuan Wang. 2004. Environmental Performance Rating and Disclosure: China's GreenWatch Program. *Journal of Environmental Management* 71 (2):123–133.

Weber, Edward P. 2003. *Bringing Society Back In: Grassroots Ecosystem Management, Accountability, and Sustainable Communities.* Cambridge, MA: MIT Press.

Weber, Ronald, and Paul Brace. 1999. States and Localities Transformed. In *American State and Local Politics: Directions for the 21st Century,* ed. Ronald E. Weber and Paul Brace. New York: Chatham House, 1–20.

Weible, Christopher M., and Paul A. Sabatier. 2009. Coalitions, Science, and Belief Change: Comparing Adversarial and Collaborative Policy Subsystems. *Policy Studies Journal: the Journal of the Policy Studies Organization* 37 (2):195–231.

Weil, David, Archon Fung, Mary Graham, and Elena Fagotto. 2006. The Effectiveness of Regulatory Disclosure Policies. *Journal of Policy Analysis and Management* 25 (1):155–181.

Weiss, Janet A., and Mary Tschirhart. 1994. Public Information Campaigns as Policy Instruments. *Journal of Policy Analysis and Management* 13:82–119.

Wilbanks, Thomas J., and Paul C. Stern. 2003. New Tools for Environmental Protection: What We Know and Need to Know. In *New Tools for Environmental Protection: Education, Information, and Voluntary Measures,* ed. Thomas Dietz and Paul C. Stern. Washington, D.C.: National Academy Press, 337–348.

Wildavsky, Aaron. 1988. *Searching for Safety.* New Brunswick, NJ: Transaction Books.

Wildavsky, Aaron. 1995. *But Is It True? A Citizen's Guide to Environmental Health and Safety Issues.* Cambridge, MA: Harvard University Press.

Williams, Bruce A., and Albert R. Matheny. 1984. Testing Theories of Social Regulation: Hazardous Waste Regulation in the American States. *Journal of Politics* 46 (2):428–459.

Williams, Bruce A., and Albert R. Matheny. 1995. *Democracy, Dialogue, and Environmental Disputes: The Contested Languages of Social Regulation.* New Haven: Yale University Press.

Williamson, Oliver E. 1985. *The Economic Institutions of Capitalism: Firms, Markets, Relational Contracting.* New York: Free Press.

Woods, Neal D. 2006. Interstate Competition and Environmental Regulation: A Test of the Race-to-the-Bottom Thesis. *Social Science Quarterly* 87 (1):174–189.

Woods, Neal D., David M. Konisky, and Ann O'M. Bowman. 2009. You Get What You Pay For: Environmental Policy and Public Health. *Publius: The Journal of Federalism* 39 (1): 95–116.

World Bank. 2000. *Greening Industry: New Roles for Communities, Markets, and Governments.* New York: Oxford University Press.

Yu, Chilik, Lawrence J. O'Toole, Jr., James Cooley, Gail Cowie, Susan Crow, and Stephanie Herbert. 1998. Policy Instruments for Reducing Toxic Releases: The Effectiveness of State Information and Enforcement Actions. *Evaluation Review* 22 (5):571–589.

Index

Accountability, 29, 35, 140, 192, 202
Abel, Troy D., 6, 41, 114, 123, 145, 180
A Citizen's Right to Know, 191
Administrative processes, 6, 14, 24, 34, 47, 129, 196
capacity of, 86, 103–105, 118, 143
professionalism and, 92–93, 111–112, 114–115
reform of, 6
Agenda-setting processes, 5–6, 11, 36, 58–60
Air pollution
characteristics of, 67, 70, 80, 126
and hazardous or toxic chemicals, 3, 16, 190, 200–201
levels of, 3, 16, 73, 78–79, 190
and state governance, 93, 102, 107, 110, 143
Alaska, 64, 75–76, 99
American Chemistry Council, 14, 121, 177, 205n14
Andrews, Richard N.L., 29
Antweiler, Werner, 26
Arkansas, 76, 80, 90, 94, 96
Ascher, William, 24
Atlas, Mark, 18, 27, 60, 180, 184, 188–190, 204n5, 217n11

Bae, Hyunhoe, 97–98
Bayer CropScience, 218n18
Beierle, Thomas C., 41, 180, 193
Bennear, Lori Synder, 178

Bhopal chemical accident
causes of, 11–12
effects of, 56–60, 199
loss of life from, 11
risk of similar accidents, 36, 199, 218n18
Blue facilities
characteristics of, 81, 94
defined, 72
Borck, Jonathan C., 197
Bosso, Christopher J., 5, 13, 189
Bouwes, Nicholass, 41, 71
Bowman, Ann O'M., 86, 97, 110
BP-owned Texas refinery, 17
Brown facilities, 55, 96, 127, 168
characteristics of, 72, 78–79, 123, 158–160, 162–163, 167, 170–174
defined, 72, 81, 155, 164
Bush administration
and environmental justice, 178
and environmental law enforcement, 218–219n19
and reform of TRI, 16, 60, 177–178, 184, 200, 205n14
and risk of terrorist attacks on chemical facilities, 16

California, 13, 59, 76, 90, 93–94, 157
Canadian National Pollutant Release Inventory, 216n6
Carbon Disclosure Project, 9
Carbon footprint information, 9
Cayford, Jerry, 193

Chemical Manufacturer's Association, 14, 121. *See also* American Chemistry Council
Chertoff, Michael, 205n14
Citizen action, 41, 137, 189
Civic environmentalism, 6
Clean Air Act, xii, 2, 15–16, 36, 178, 188, 200–201
    Amendments of 1990, 6, 190
    and cap-and-trade program for acid rain, 6
    economic benefits of, 3
    national ambient air quality standards under, 210n7
Clean Water Act, xii, 2, 178, 207n3
Climate change, 9, 23–24, 178–79, 216n5
Climate Counts, 9
Coase, Ronald, 37–39
Code of Federal Regulations and TRI, 61–62
Coglianese, Cary, 7, 42, 178, 180, 194, 197–198
Cohen, Mark, 26, 89
Collective action dilemmas, 38–39, 87, 96, 117, 183
Columbus Pulp and Paper, 28
Command-and-control regulation, xii, 3–7, 15, 39, 47, 66
    achievements of, 3
    alternatives to, 6–7, 15, 23, 178
    shortcomings of, 4
Communities, 55, 63, 147, 191, 200
    access to information by, 13, 138, 180, 197
    capacity of, 44, 120, 161, 174–175, 182
    chemical pollution in, 32
    impacts of chemicals on, 1–2, 7, 10–11
    and information disclosure, 25, 33, 40, 134, 182
    leadership within, 45, 168, 182
    and local industries, 17, 37, 59, 81, 180, 191
    poor and minority populations in, 10, 177
    pressure from, 27, 38, 42, 138, 155, 166
    resources within, 44
    right-to-know in, 31, 59
    risk to, 11, 26, 54, 62, 98, 162, 181
    and TRI, 10, 18–19, 61, 83, 129, 148, 183, 195
Community Air Screening How-To Manual, A Step-by-Step Guide to Using Risk-Based Screening to Identify Priorities for Improving Outdoor Air Quality, 217n13
Comprehensive Environmental Response, Compensation and Liability Act. *See* Superfund program
Connecticut, 13, 76, 90–91, 94, 157
Corporations. *See* Industry
Cost(s). *See also* Economic(s); Incentives
    acceptable risks and, 146
    of compliance, 3–5, 10, 18, 20, 38–39, 44, 58, 188
    of environmental protection,5, 177, 184, 191, 195, 202
    of pollution control, 19, 141–142, 164, 167, 169, 173
Cytec Industries, 53

Daley, Dorothy M., 193
Davies, J. Clarence III, 4, 31
Delaware, 76, 90–91, 94
de Marchi, Scott, 27, 69, 88, 184
Democratic Party, 103
Department of Homeland Security, 16, 199, 205n14
Dietz, Thomas, 5, 7, 178, 193
Douglas, Mary, 30, 40
Dow Chemical Company, 204n10
Downs, Anthony, 38
Drinking Water Quality, 2–3, 8, 13, 30, 56, 200, 207n3, 219n19
Duke Energy, 1
DuPont, 1
Durant, Robert F., 4–5, 7

Eastman Kodak, 53
Economic incentives and
  environmental policy, xii, 4–6, 39.
  See also Incentives; Economics
Economics, 44, 116. *See also* Costs;
  Incentives
Eisner, Marc Allen, 4, 6, 14, 35, 62,
  165, 178, 198
Emergency Planning and Community
  Right to Know Act, 12, 36, 57
  and emergency planning, 123, 136,
  166, 171, 185
Enforcement dilemma, 47, 183
Enron, 8
Environmental Defense Fund, 54,
  187–188, 217n8
Environmental equity, 10, 19, 98,
  202
Environmental federalism, 65, 85
Environmental health and safety, 163
  social movements and, 14
  staff in industrial facilities, 127–128
Environmental laggards, 19, 75, 84,
  153–175, 182
Environmental leaders, 134, 153–175
  determinants of, 117
  differences between laggards and,
  19, 153–175
  facilities with, 75
Environmental management systems
  (EMSs), 65, 69, 127, 131, 164
  improving, 18, 122, 197–198
  influence of, 165
  integration of with corporate
  decisions, 14, 26, 55, 130, 167
  performance of, 140–141, 148, 166
  requirements of, 85, 130, 190
  results from, 168
Environmental performance
  and corporate actions, xi, 1, 4, 16,
  25–26, 67, 85–88, 158, 167, 172,
  180
  and economic incentives/impacts,
  43–44, 102, 146, 191, 196
  and employee incentives, 164, 174
  and EMSs, 130, 168
  facility interaction and, 171, 182

guidelines and regulations and, 14,
  20, 25
and information disclosure, 2, 7, 17,
  20, 28, 31, 39, 42, 107, 197
measures of, 20, 50, 71–72, 89,
  103, 125
media coverage of, 139
and public health, 18, 37, 180, 185
research directed at, 32, 34, 91, 153
state performance and, 21, 91,
  93–94, 96–99, 105, 107, 117, 186
transparency and, 173, 190
Environmental performance dilemma.
  See Performance dilemma
Environmental policy. *See also*
  Environmental Protection Agency;
  Toxics Release Inventory
  public awareness of, 5
  reform agenda for, xii, 4–7, 179,
  183, 201–202
  use of science in, 24
Environmental Protection Agency.
  See also Toxics Release Inventory
  administrative rulemaking and, 5,
  16, 54, 67–68, 106, 177–178, 200
  air quality and, 3, 9, 15, 26–28, 93,
  190, 200–201
  balancing statutory goals and costs
  at, 10, 84
  Community Action for a Renewed
  Environment (CARE) program,
  217n13
  enforcement and compliance, 2–3,
  41, 53–54, 60–65, 69–71, 89, 113,
  128, 139, 149, 179, 181–182
  environmental justice and, 177
  Office of Pollution Prevention and
  Toxic Substances, 217n13
  National Center for Environmental
  Research and Quality Assurance,
  213n12
  risk assessment by, 10, 12, 15–17,
  26–29, 37, 55–56, 88, 93, 157,
  159, 184
  science at, 98, 106, 178, 187
  Superfund and, 193
  water quality and, 3, 200

Environmental Working Group, 207n5

Erikson, Robert S., 45, 93, 103, 116

Evaluation
of alternative policies, 106, 132, 202
approaches to, 121
cost-benefit analysis in, 164, 167,
169, 173
data assessment in, 7, 20, 66–67,
84, 120, 155–156, 179, 182, 201

Facebook, 191

Federal Insecticide, Fungicide and
Rodenticide Act, 2

Financial disclosure requirements,
8–9, 57–58

Fiorino, Daniel J., 2, 4–7, 13, 39, 47,
62, 126, 140, 148, 171, 178, 192,
194

Florio, James, 14, 36, 57–58, 60–62,
121, 123, 149

Food Quality Protection Act, 6, 8
and information disclosure, 30

Friends of the Earth, 103

Fung, Archon, 8–9, 26, 41, 120, 131,
140, 180

Game theory, 19, 39

General Electric, 1

Gerber, Brian J, 45, 86–87, 116

Global Reporting Initiative, 216n6

Gormley, William T., 9–10, 35–36

Government Accountability Office
(GAO), 180, 200
on environmental justice, 177
on release levels, 69
study of TRI, 188–190, 192–193
on toxic chemical policy
implementation, 205n12

Graham, Mary, 8–9, 18, 26, 49, 54,
71, 180, 184

Grant, Don, 26, 85–87, 97, 120, 126,
131, 180, 183, 193

Green facilities
characteristics of, 78–79, 81, 91, 94,
96, 123, 131, 156–158, 160–166,
168, 170–175
defined, 72, 155

Greenhouse gases, 1, 9, 179, 204,
216. *See also* Climate change

Guber, Deborah Lynn, 203–204n5

Gunningham, Neil, 26, 126, 198

Hadden, Susan G., 10–11, 13, 18,
180, 191

Hall, Bob, 93, 105

Hamilton, James T., 6, 14, 18, 27,
36–37, 42, 47, 63, 66, 69, 88, 120,
136–137, 172, 180, 184

Harrison, Kathryn, 7, 26

Harvard's Kennedy School of
Government, 9

Hassur, Steven, 41, 71

Health risks, human, 50, 75, 187
acceptable levels of, 181, 201
of air pollution, 16, 26, 37, 70, 79,
200
analysis of, 27, 49, 71–72, 135,
185
chronic, 129
citizen understanding of, 186, 188,
196, 199
of hazardous waste, 11, 182, 190,
200
management of, 192
reduction of, 56, 66, 184, 198
of toxic chemicals, 2, 10, 19, 24,
25, 31, 34, 58, 62, 70, 133, 162,
181, 184–185, 200–201
of water pollution, 66–68, 70, 200,
207n3, 219n19

Healy, Robert, 29

Herb, Jeanne, 10, 24, 31, 41, 180

Helms, Susan, 10, 24, 31, 41, 180

Hybrid environmental policies, xii,
196–198, 201–202

Illinois, 76, 90, 94

Incentives. *See also* Economics;
Costs
for employees, 130, 168
for facilities, 147, 168, 184, 194
market, xii, 5–6
and the performance dilemma, 116
to reduce pollution, 39, 119, 191

Industry
 behavior of, xi, 1, 20, 25, 37, 46,
  48, 55, 81, 98, 103, 120, 138, 145,
  155
 citizens and, 37, 46, 140, 198
 environmental performance of,
  xi–xii, 4, 16, 18, 25–26, 78, 93,
  125–126, 156, 180–181
 greening of, 1, 28, 39, 47, 84, 93,
  153
 social responsibility of, 1, 43, 55,
  81, 198
 state regulation of, 103, 106, 114
 types of, 75, 125
Information acquisition, 40, 192
 cost of, 38–39
Information asymmetry, 39, 116
Information deficiency, 39
Information disclosure
 burdens of, 128, 177, 184, 191,
  195–196, 202
 capacity building and, 19, 42–43,
  45–48, 87, 128
 and carbon footprints, 9
 challenges of, 34, 187–189
 and community involvement, 13, 14,
  40, 180, 183–188, 191–193
 effectiveness of, xii, 3, 29–31, 39,
  51, 178–179, 183, 201
 and greenhouse gases, 1, 9, 179,
  204n6, 204n8, 216n5 (*see also*
  Climate change)
 impacts of, xii, 7, 16, 32, 36,
  85–87, 165
 limitations of, 17, 189
 mandatory, xii, 14, 66, 197–198,
  204n2
 moral responsibility for, 35
 potential for, xii, 2, 21, 56, 178
 process of, 138–139, 155
 programs of, 2, 7–11, 14–15, 39,
  47, 190
 state use of, 6, 107, 137, 186,
  195
 transaction costs of, 37, 39
 and TRI, xii, 33, 35, 41, 47, 55–56,
  176, 180, 197, 201–202

 types of policies of, 8–11, 36, 42,
  57–59, 114, 195–202
International Organization for
 Standardization, 14, 85
 and ISO certification, 85, 130, 145,
  165, 168, 175, 190
Information provision, 203. *See also*
 Information disclosure
Ingram, Helen, 42
Invista, 54
Issue framing. *See* Agenda setting

Jackson, Lisa P., 178
Jenkins-Smith, Hank C., 206n7
Jensen, Michael J., 10, 24, 31, 41,
 180
John, DeWitt, 6–7
Johnson Controls, 1
Jones, Andrew W., 26, 85, 120, 126,
 131, 180, 183, 193

Kagen, Robert A., 26, 126, 198
Kamieniecki, Sheldon, 5, 58, 62
Kansas, 76, 90, 94, 107
Kartez, Jack D., 26, 41, 121, 131,
 137, 180, 189
Keller, Ann Campbell, 24
Kerr, Mary Lee, 93, 105
Kingdon, John, 11, 36, 58–60
Klyza, Christopher McGrory, 6,
 179
Konar, Shameek, 26, 89
Konisky, David M., 85, 97, 102,
 110
Kraft, Michael E., 4–7, 12, 24,
 35–36, 41, 58, 62, 114, 118, 123,
 145, 179, 200

Laggards, in environmental
 performance, 19–21, 134, 153–
 176, 182, 194
Lautenberg, Frank R., 178
Layzer, Judith A., 7
Leaders, in environmental
 performance, 19–21, 134,
 153–176, 182, 194
Lester, James P., 86–87, 97, 112

Local Emergency Planning
   Committees (LEPCs)
and chemical management, 129
and community leadership, 45
interactions with facilities, 136, 140,
   171
response rates of, 123–124
states and, 130
use of TRI, 133, 196
Louisiana, 53, 76, 80
Lubell, Mark, 7
Lynn, Francis M., 26, 41, 121, 131,
   137, 180, 189
Lyon, Thomas P., 42, 131, 136, 140,
   143, 145, 198

Maine, 83–84, 90–91
Market incentives. *See* Costs;
   Economics; Incentives
Maryland, 76, 90, 98, 189
Massachusetts, 65, 76, 90, 93–94,
   106, 114, 157
Massachusetts Public Interest
   Research Group (MASSPIRG), 115
Maxwell, John W., 42, 131, 136,
   140, 143, 145, 198
Mazmanian, Daniel A., 5–6, 39, 43,
   44, 54, 63, 84, 88, 118, 153, 179,
   206n5
Mazurek, Janice, 4
McIver, John, 45, 93, 103, 116
Miller, Catherine, 27, 49, 63, 69, 184
Monsanto, 53, 157
Morgenstern, Richard D., 7, 178,
   198
MySpace, 191

Nash, Jennifer, 7, 42, 178, 180, 194,
   198
Natan, Thomas E., Jr., 27, 49, 63,
   69, 184
National Academy of Public
   Administration, 201
National Conference of State
   Legislatures (NCSL), 107
National Emissions Inventory, 188,
   190

National Environmental Performance
   Track program, 194
National-Scale Air Toxics Assessment
   (NATA), 201, 219n20
National Wildlife Federation, 103
Nelson, Kimberly, 178
Nevada, 64, 76, 90–91, 93–94, 99
New Hampshire, 76, 90, 94, 157
New Jersey, 14, 59, 65, 76, 90, 94,
   107
New Mexico, 76, 89–90, 94, 96, 99

Obama, Barack
and National Environmental
   Performance Track program, 194
and TRI program, 60, 178, 205n16
Occupational Safety and Health
   Administration (OSHA), 12, 59
Office of Management and Budget
   (OMB), 216n4
Ohio, 75–76, 84, 89–90, 94, 145
O'Leary, Rosemary, 4–5, 7
Olmstead, Sheila M., 6
Olson, Mancur, 38
OMB Watch, 54
Omnova Solutions, 28–29
O'Rourke, Dara, 26, 41, 44, 106,
   131
O'Toole, Randall, 117

Paperwork Reduction Act, 216n4
Pennsylvania, 76, 90, 94
Performance dilemma, 81, 103, 116,
   180, 186, 206n5, 211n10
defined, 19, 39, 46–47, 86–87
escape from, 96, 117–118, 127,
   148, 181, 198–199
factors affecting, 129, 130
and performance synergy, 46–47,
   198
and TRI, xii, 121–122, 133
Persistent, bioaccumulative, and toxic
   (PBT) chemicals, 12, 14, 16, 56,
   62
Pizer, William A., 7, 198
Policy design, 19
Policy entrepreneurs, 58, 59

Policy implementation, 2, 4, 7, 19, 153
of EPCRA, 121
states and, 89, 183
of Superfund, 53, 59
of TRI, 7, 19, 37
Policy learning, 47, 192, 194, 206n7
Policy streams, 11, 58–61
Pollution prevention, 40, 45, 54, 199
administrative capacity and, 103–106
assistance programs in, 182, 186, 195–197
implementation of, 89
integration of, 93, 109, 111–115
legislation and, 108, 110, 117–118
and TRI, 142–143, 148, 186
Pollution Prevention Act (1990), 12, 20, 61, 134–135
Polychlorinated biphenyls (PCBs), 16, 67
Popp, David, 97–98
Portland Press Herald, 83
Portney, Paul R., 4, 203
Potoski, Matthew, 6–7, 39, 47, 55, 66, 85–87, 102, 118, 130, 153, 165, 197, 206n5
Prakash, Aseem, 6–7, 39, 47, 55, 66, 85, 87, 118, 130, 153, 165, 197, 206n5
Press, Daniel, 39, 43, 44, 54, 63, 84, 88, 153, 183, 206n5
Primary metals, 3, 75, 77, 79, 125–126
Principal-agent theory, 45–46
Prisoner's dilemma, 39, 46, 117–118, 206n4
Problem stream, 11, 58
Production related waste (PRW), 15, 71, 208n15, 210n7
Proposition 65, the Safe Drinking Water and Toxic Enforcement Act, 13
Public participation
in EPA decision making, 193
in policy evaluation, 151
in policymaking, 14, 193

Rabe, Barry G., 6, 45, 65, 85, 112, 212–213n16
Race-to-the-bottom thesis, 85, 102–103
Reagan administration, 12, 60, 207
Regulation dilemma, 38–39, 46, 66, 130
and industrial performance, 86
and safety enforcement, 86
Regulatory compliance, 3, 4, 46, 81, 84, 143–144
flexibility in, 5
going beyond, 122, 134, 198
improvement in, 140, 147, 154, 194–195
Resource Conservation and Recovery Act, 2, 36
Responsible Care initiative, 14, 63, 157
Rhode Island, 76, 91, 93–94
Right-to-know legislation
history of, 12–14, 35–36, 57
principles of, 11, 58, 65–66, 191, 199, 202
in the states, 13, 59–60, 113, 132, 183
Ringquist, Evan J., 14, 35, 45, 86–87, 93, 97–98, 102, 105, 110–111
Risk perception and communication, 14, 29–31, 180, 185, 188, 217n12
Risk Screening Environmental Indicators (RSEI) model, 71, 123, 188
described, 26–27
development of, 97
importance of, 50, 88, 216
and measurement of toxic chemical pollutants, 98
and risk changes, 91
risk levels measured by, 49, 55, 70–72, 185, 187
second generation model, 196
RSEI. *See* Risk Screening Environmental Indicators

Sabatier Paul A., 206n7
Safe Drinking Water Act, 2, 8, 13
  oversight of, 200, 207n3
  requirements under, 30
Sarbanes-Oxley Act 2002, 8
Scheberle, Denise, 45
Schneider, Anne Larason, 42
Schoenbrod, David, 7, 179
Scholz, John, 38–39, 46–47, 86–87, 183
S.C. Johnson Company, 1
Securities Exchange Act, 57
Securities Exchange Commission (SEC), and greenhouse gas financial risks, 9
Shapiro, Mark D., 41, 71
Shapiro, Michael, 68
Sierra Club, 103
Sigman, Hillary, 68, 207n10
Slovic, Paul, 13, 30, 40
Smith, Eric R.A. N., 204n5
Source reduction, 12, 61, 72
  effectiveness of, 143
  opportunities for, 134–135, 141
  strategies of, 106, 125, 141
  and TRI, 119, 132, 142, 148
Sousa, David, 6, 179
South Carolina, 76, 91, 94, 96
Stavins, Robert N., 4
State(s), 13, 59, 137, 182
  comparative analysis of, 21, 29, 50, 64–65, 75–76, 81–99, 112, 157–158
  environmental performance in, 3, 20, 27, 114–118, 124, 153
  and LEPCs, 130
  policymaking and, 3, 5–6, 13, 45, 59–60, 85, 102–109, 134, 145, 177, 190, 194–196, 202
  progress in pollution reduction in, 55, 110, 113, 115
  TRI facilities in, 18, 25, 34, 61, 74, 160, 176
  use of TRI Explorer and Envirofacts in, 192–193
Steelman, Toddi, 29

Stephan, Mark, 6, 26, 32, 41, 114, 123, 145
Stern, Paul C., 5, 7, 178, 193
Superfund Amendments and Reauthorization Act (1986), 12, 53, 178
Superfund Program, 2, 12
  citizen groups and, 193
  reauthorization and broadening of, 13, 57, 59, 63
Sustainable development, 4–5, 19, 73, 88, 179

Tennessee, 76, 89, 91, 94
Teske, Paul, 45, 85–87, 116
Thornton, Dorothy, 26, 126, 198
Total Production Related Waste (TPRW), 15
Toxic chemicals
  accidents involving, 11–12, 36, 56–59, 129, 138
  air quality and, 70, 72, 79, 123, 190
  alternatives to, 119, 146
  disposal of, 68, 71, 182
  emission of, 40, 48, 53, 138, 153
  health effects of, 58–59, 62, 66–67, 70, 98, 133, 162, 190, 196
  information about, 10, 12, 69, 181, 187–188
  management of, 13, 19, 21, 58, 106, 119–122, 125, 136, 153, 166, 184
  release of, 3, 15–18, 25, 34, 54, 56, 62–63, 71, 74, 78–80, 83, 139
  right to know about, 60–61, 191, 193
  risks of, 24, 55, 57, 158, 199
  water quality and, 67, 70, 123
Toxicological effects, 71–72, 123
Toxics Release Inventory (TRI)
  air releases in, 70, 73, 123, 147
  burden reduction and, 177–178, 205n16, 215n1, 216n4
  citizen involvement with, 10, 24, 40, 44, 196
  and community right to know, 26, 31, 202

data quality of, 17, 61–62, 65, 71, 75
effects of, xii, 14, 20, 32, 46, 83, 122, 143
and environmental performance, xii, 34, 47, 50, 54, 80, 93, 198
facility perceptions of, 26–28, 43, 129, 131, 170
federal officials and, 36, 85, 185–186
history of, 7, 11–12, 55–60, 157, 199
impacts of, xi, 37, 55, 74, 142, 145, 180
impacts on environmental management, 172–176
limitations of, 10, 50, 71–72, 150, 183–184
local officials and, 51, 139, 148, 185
media coverage of, 26, 43, 50, 136, 157, 172
policymakers and, 55, 81, 201
state officials and, 51, 85, 87–89, 96, 99, 113, 139, 148, 185–186
transaction costs of, 38, 44, 64
trends in over time, 15–19, 23, 49, 66–67, 154, 181–182
use of data from, 132–135, 140–142, 179–180, 188, 190, 193, 195
Toxic Substances Control Act (1976), 2, 207–208n10
Toxics Use Reduction Institute (TURI), 106, 115
Toxic Use Reduction Act (TURA), 65, 106, 115
Transparency, 1, 29, 38, 57, 192, 202
corporate, 57
Policy Project at Harvard, 9
in release of toxic chemicals, 53, 140
in rulemaking, 14, 66
TRI Explorer, 193
TRI-ME, 65
Twitter, 191, 217

Underground injection, 3, 68
Union Carbide, 204
Bhopal accident and, 11, 59–60
plant in West Virginia, 11, 36, 58
USA Today, 137, 187
U.S. Environmental Protection Agency (U.S. EPA). *See* Environmental Protection Agency
Utah, 76, 91, 94

Vermont, 76, 89, 91, 93–94, 99, 157
Vig, Norman J., 6, 12, 24, 179, 200
Vogel, David, 198
Voluntary environmental programs, 6, 9, 47, 197–198, 201
33/50 program, 54, 191
of emissions reduction, 53, 66
of pollution prevention, 54, 186
Project XL, 191

Wal-Mart, 1
Washington State, 76, 91, 94
Weil, David L., 8–9, 120, 140, 180
Weimer, David, 9–10
Wilcoxen, Peter, 97–98
Wildavsky, Aaron, 30, 40
Wisconsin Department of Natural Resources, and air screening models, 216–217n7
Wisconsin's Green Tier Program, 194, 218n17
Woods, Neal D., 85–86, 87, 97, 102, 110
Workers and Community Right to Know Act, 59
World Business Council for Sustainable Development, 216n6
Worsham, Jeff, 14, 35.
Wright, Gerald, 45, 93, 103, 116

Yellow facilities
characteristics of, 92, 94, 198
defined, 72, 81
Yu, Chilik, 26, 85, 87, 106–107, 117

American and Comparative Environmental Policy
Sheldon Kamieniecki and Michael E. Kraft, series editors

Russell J. Dalton, Paula Garb, Nicholas P. Lovrich, John C. Pierce, and John M. Whiteley, *Critical Masses: Citizens, Nuclear Weapons Production, and Environmental Destruction in the United States and Russia*

Daniel A. Mazmanian and Michael E. Kraft, editors, *Toward Sustainable Communities: Transition and Transformations in Environmental Policy*

Elizabeth R. DeSombre, *Domestic Sources of International Environmental Policy: Industry, Environmentalists, and U.S. Power*

Kate O'Neill, *Waste Trading among Rich Nations: Building a New Theory of Environmental Regulation*

Joachim Blatter and Helen Ingram, editors, *Reflections on Water: New Approaches to Transboundary Conflicts and Cooperation*

Paul F. Steinberg, *Environmental Leadership in Developing Countries: Transnational Relations and Biodiversity Policy in Costa Rica and Bolivia*

Uday Desai, editor, *Environmental Politics and Policy in Industrialized Countries*

Kent Portney, *Taking Sustainable Cities Seriously: Economic Development, the Environment, and Quality of Life in American Cities*

Edward P. Weber, *Bringing Society Back In: Grassroots Ecosystem Management, Accountability, and Sustainable Communities*

Norman J. Vig and Michael G. Faure, editors, *Green Giants? Environmental Policies of the United States and the European Union*

Robert F. Durant, Daniel J. Fiorino, and Rosemary O'Leary, editors, *Environmental Governance Reconsidered: Challenges, Choices, and Opportunities*

Paul A. Sabatier, Will Focht, Mark Lubell, Zev Trachtenberg, Arnold Vedlitz, and Marty Matlock, editors, *Swimming Upstream: Collaborative Approaches to Watershed Management*

Sally K. Fairfax, Lauren Gwin, Mary Ann King, Leigh S. Raymond, and Laura Watt, *Buying Nature: The Limits of Land Acquisition as a Conservation Strategy, 1780–2004*

Steven Cohen, Sheldon Kamieniecki, and Matthew A. Cahn, *Strategic Planning in Environmental Regulation: A Policy Approach That Works*

Michael E. Kraft and Sheldon Kamieniecki, editors, *Business and Environmental Policy: Corporate Interests in the American Political System*

Joseph F. C. DiMento and Pamela Doughman, editors, *Climate Change: What It Means for Us, Our Children, and Our Grandchildren*

Christopher McGrory Klyza and David J. Sousa, *American Environmental Policy, 1990–2006: Beyond Gridlock*

John M. Whiteley, Helen Ingram, and Richard Perry, editors, *Water, Place, and Equity*

Judith A. Layzer, *Natural Experiments: Ecosystem-Based Management and the Environment*

Daniel A. Mazmanian and Michael E. Kraft, editors, *Toward Sustainable Communities: Transition and Transformations in Environmental Policy*, 2d edition

Henrik Selin and Stacy D. VanDeveer, editors, *Changing Climates in North American Politics: Institutions, Policymaking, and Multilevel Governance*

Megan Mullin, *Governing the Tap: Special District Governance and the New Local Politics of Water*

David M. Driesen, editor, *Economic Thought and U.S. Climate Change Policy*

Kathryn Harrison and Lisa McIntosh Sundstrom, editors, *Global Commons, Domestic Decisions: The Comparative Politics of Climate Change*

William Ascher, Toddi Steelman, and Robert Healy, *Knowledge and Environmental Policy: Re-Imagining the Boundaries of Science and Politics*

Michael E. Kraft, Mark Stephan, and Troy D. Abel, *Coming Clean: Information Disclosure and Environmental Performance*